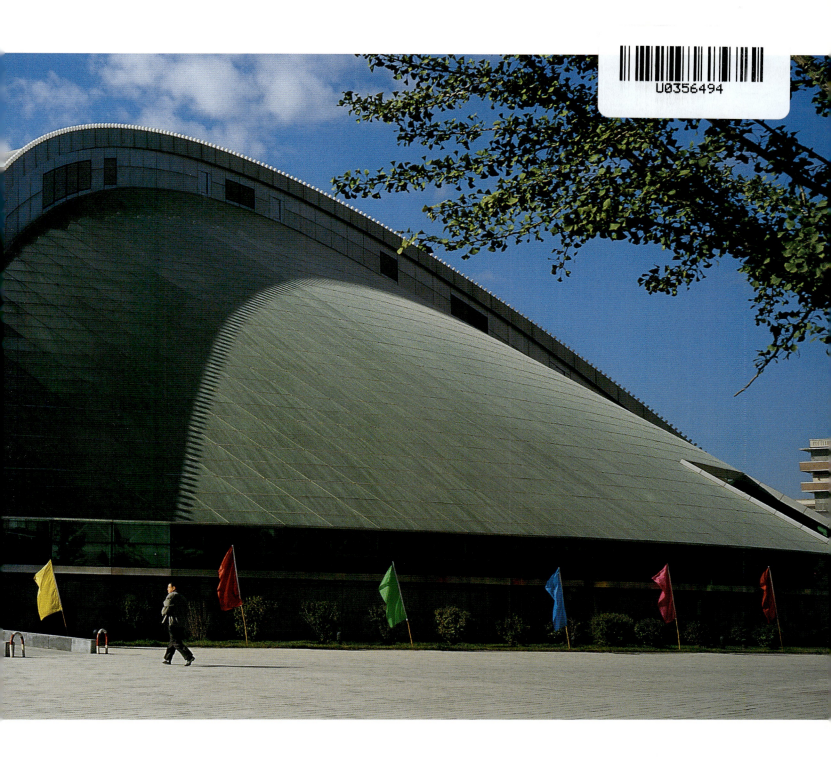

如欲了解详情，请联络以下单位：

亨特建材（上海）有限公司
电话：86-21-6442-9999
电子邮件：magnate@hunterdouglas.sh.cn

亨特建材（深圳）有限公司
电话：86-755-8369-9600
电子邮件：salesap@hunterdouglas.com.cn

亨特建材（北京）有限公司
电话：86-10-6788-9900
电子邮件：ap@hunterdouglas.bj.cn

亨特建筑构件（厦门）有限公司
电话：86-592-651-2811
电子邮件：hdxm@hunterdouglas.com.cn

©2005 Hunter Douglas Group版权注册 ® Hunter Douglas Group注册商标

HunterDouglas

WINDOW COVERINGS | CEILINGS | SOLAR CONTRCL | FACADES

助 您 开 通 信 息 渠 道

connecting people _projects _products

The McGraw·Hill Companies

引领信息资讯潮流
称雄建筑信息市场

麦格劳－希尔建筑信息公司 (McGraw Hill Construction) 将建筑业信息与资讯完美地结合在一起，领跑业界百余年，不断树立新标准。

我们享有盛誉的**道奇(Dodge)**、**斯维茨(Sweets)**、**建筑实录(Architectural Record)**、**工程新闻记录(Engineering News-Record)**、地区出版物，以及行业门户网站 www.construction.com 构成我们强大的信息资源网。所有这一切为全球3万4千亿美元的建筑市场赢得了超过一百万家客户。

拓展海外建筑市场

最佳连接

麦格劳－希尔建筑信息公司凭借其丰富的产品与服务，不但能够帮助您建立更强大、更广泛的业务联系，更可使美国或世界其他地方的决策者迅速了解您的情况，使您的工作更加简便、迅速。

明智决策

从行业新闻及趋势，到项目信息和产品信息，麦格劳-希尔建筑信息公司为中国建筑企业带来了成功开拓海外市场的钥匙：必要的工具，资源及措施。拥有这一切，无论形势如何变幻莫测，您都可以了然于胸，做出更明智的决定。

成就非凡

在建筑业，打造关系胜于一切。一个多世纪以来，麦格劳-希尔建筑信息公司成功运用其丰富经验和资源，帮助不计其数的企业一步步成长为业界领袖。让我们来帮助您，在美国及海外建筑市场打下一片天地。请现在就与我们联系，告知我们您的需求。

联系人：许敏达
业务拓展高级主管
Minda Xu
Senior Director, Business Development
McGraw-Hill Construction

Two Penn Plaza, 9th Fl.
New York, NY 10121
212-904-3519 Tel
212-904-4460 Fax
minda_xu@mcgraw-hill.com

McGraw_Hill CONSTRUCTION

Dodge
Sweets
Architectural Record
ENR
Regional Publications

Find us online at www.construction.com

ARCHITECTURAL RECORD

EDITOR IN CHIEF	Robert Ivy, FAIA, *rivy@mcgraw-hill.com*
MANAGING EDITOR	Beth Broome, *elisabeth_broome@mcgraw-hill.com*
DESIGN DIRECTOR	Anna Egger-Schlesinger, *schlesin@mcgraw-hill.com*
DEPUTY EDITORS	Clifford Pearson, *pearsonc@mcgraw-hill.com*
	Suzanne Stephens, *suzanne_stephens@mcgraw-hill.com*
	Charles Linn, FAIA, Profession and Industry, *linnc@mcgraw-hill.com*
SENIOR EDITORS	Sarah Amelar, *sarah_amelar@mcgraw-hill.com*
	Sara Hart, *sara_hart@mcgraw-hill.com*
	Deborah Snoonian, P.E., *deborah_snoonian@mcgraw-hill.com*
	William Weathersby, Jr., *bill_weathersby@mcgraw-hill.com*
	Jane F. Kolleeny, *jane_kolleeny@mcgraw-hill.com*
PRODUCTS EDITOR	Rita F. Catinella, *rita_catinella@mcgraw-hill.com*
NEWS EDITOR	Sam Lubell, *sam_lubell@mcgraw-hill.com*
DEPUTY ART DIRECTOR	Kristofer E. Rabasca, *kris_rabasca@mcgraw-hill.com*
ASSOCIATE ART DIRECTOR	Clara Huang, *clara_huang@mcgraw-hill.com*
PRODUCTION MANAGER	Juan Ramos, *juan_ramos@mcgraw-hill.com*
WEB EDITOR	Randi Greenberg, *randi_greenberg@mcgraw-hill.com*
WEB DESIGN	Susannah Shepherd, *susannah_shepherd@mcgraw-hill.com*
WEB PRODUCTION	Laurie Meisel, *laurie_meisel@mcgraw-hill.com*
EDITORIAL SUPPORT	Linda Ransey, *linda_ransey@mcgraw-hill.com*
ILLUSTRATOR	I-Ni Chen
EDITOR AT LARGE	James S. Russell, *AIA, james_russell@mcgraw-hill.com*
CONTRIBUTING EDITORS	Raul Barreneche, Robert Campbell, FAIA, Andrea Oppenheimer Dean, Francis Duffy, Lisa Findley, Blair Kamin, Elizabeth Harrison Kubany, Nancy Levinson, Thomas Mellins, Robert Murray, Sheri Olson, AIA, Nancy Solomon, AIA, Michael Sorkin, Michael Speaks, Tom Vonier, AIA
SPECIAL INTERNATIONAL CORRESPONDENT	Naomi R. Pollock, AIA
INTINTERNATIONAL CORRESPONDENTS	David Cohn, Claire Downey, Tracy Metz
GROUP PUBLISHER	James H. McGraw IV, *jay_mcgraw@mcgraw-hill.com*
VP, ASSOCIATE PUBLISHER	Laura Viscusi, *laura_viscusi@mcgraw-hill.com*
VP, EVENTS AND BUSINESS DEVELOPMENT	David Johnson, *dave_johnson@mcgraw-hill.com*
VP, GROUP EDITORIAL DIRECTOR	Robert Ivy, FAIA, *rivy@mcgraw-hill.com*
GROUP DESIGN DIRECTOR	Anna Egger-Schlesinger, *schlesin@mcgraw-hill.com*
DIRECTOR, CIRCULATION	Maurice Persiani, *maurice_persiani@mcgraw-hill.com*
	Brian McGann, *brian_mcgann@mcgraw-hill.com*
DIRECTOR, MULTIMEDIA DESIGN & PRODUCTION	Susan Valentini, *susan_valentini@mcgraw-hill.com*
DIRECTOR, FINANCE	Ike Chong, *ike_chong@mcgraw-hill.com*
PRESIDENT, MCGRAW-HILL CONSTRUCTION	Norbert W. Young Jr., FAIA

EDITORIAL OFFICES: 212/904-2594. Editorial fax: 212/904-4256. E-mail: rivy@mcgraw-hill.com. Two Penn Plaza, New York, N.Y. 10121-2298. **WEB SITE:** www.architecturalrecord.com. **SUBSCRIBER SERVICE:** 877/876-8093 (U.S. only). 609/426-7046 (outside the U.S.). Subscriber fax: 609/426-7087. E-mail: p64ords@mcgraw-hill.com. AIA members must contact the AIA for address changes on their subscriptions. 800/242-3837. E-mail: members@aia.org. **INQUIRIES AND SUBMISSIONS:** Letters, Robert Ivy; Practice, Charles Linn; Books, Clifford Pearson; Record Houses and Interiors, Sarah Amelar; Products, Rita Catinella; Lighting, William Weathersby, Jr.; Web Editorial, Randi Greenberg

McGraw_Hill CONSTRUCTION — The McGraw·Hill Companies

This Yearbook is published by China Architecture & Building Press with content provided by McGraw-Hill Construction. All rights reserved. Reproduction in any manner, in whole or in part, without prior written permission of The McGraw-Hill Companies, Inc. and China Architecture & Building Press is expressly prohibited.

《建筑实录年鉴》由中国建筑工业出版社出版，麦格劳希尔提供内容。版权所有，未经事先取得中国建筑工业出版社和麦格劳希尔有限总公司的书面同意，明确禁止以任何形式整体或部分重新出版本书。

建筑实录 年鉴 VOL.2/2005

主编 EDITORS IN CHIEF
Robert Ivy, FAIA, *rivy@mcgraw-hill.com*
赵晨 *zhaochen@china-abp.com.cn*

编辑 EDITORS
Clifford Pearson, *pearsonc@mcgraw-hill.com*
率琦 *shuaiqi@china-abp.com.cn*
戚琳琳 *qll@china-abp.com.cn*

新闻编辑 NEWS EDITOR
Sam Lubell, *sam_lubell@mcgraw-hill.com*

撰稿人 CONTRIBUTORS
Jen Lin-Liu, Dan Elsea, Shirley Chang

美术编辑 DESIGN AND PRODUCTION
Anna Egger-Schlesinger, *schlesin@mcgraw-hill.com*
Kristofer E. Rabasca, *kris_rabasca@mcgraw-hill.com*
Clara Huang, *clara_huang@mcgraw-hill.com*
Juan Ramos, *juan_ramos@mcgraw-hill.com*
蔡宏生 *chs@china-abp.com.cn*
许萍 *picachuxu@163.com*

特约顾问 SPECIAL CONSULTANT
支文军 *ta_zwj@163.com*
王伯扬

翻译 TRANSLATORS
徐洁 *jjiexu@citiz.net*
蔡瑜 *carolfish117@yahoo.com.cn*
蒋妙菲 *lh79@vip.sina.com*

中文制作 PRODUCTION, CHINA EDITION
同济大学《时代建筑》杂志工作室

中文版合作出版人 ASSOCIATE PUBLISHER, CHINA EDITION
Minda Xu, *minda_xu@mcgraw-hill.com*
张惠珍 *zhz@china-abp.com.cn*

市场营销 MARKETING MANAGER
Lulu An, *lulu_an@mcgraw-hill.com*
白玉美 *bym@china-abp.com.cn*

广告制作经理 MANAGER, ADVERTISING PRODUCTION
Stephen R. Weiss, *stephen_weiss@mcgraw-hill.com*

印刷/制作 MANUFACTURING/PRODUCTION
Michael Vincent, *michael_vincent@mcgraw-hill.com*
Kathleen Lavelle, *kathleen_lavelle@mcgraw-hill.com*
Carolynn Kutz, *carolynn_kutz@mcgraw-hill.com*
王雁宾 *wyb@china-abp.com.cn*

著作权合同登记图字：01-2005-1956 号

图书在版编目（CIP）数据
建筑实录年鉴 VOL.2/2005. —北京：
中国建筑工业出版社，2005
ISBN 7-112-07560-2
Ⅰ. 建⋯ Ⅱ. 建⋯ Ⅲ. 建筑实录 -2005- 年鉴 Ⅵ. TU206-54
中国版本图书馆 CIP 数据核字（2005）第 091258 号

建筑实录年鉴 VOL.2/2005

中国建筑工业出版社出版、发行（北京西郊百万庄）
新华书店经销
上海当纳利印刷有限公司印刷
开本：880 × 1230 毫米 1/16 印张：4¼ 字数：180 千字
2005 年 8 月第一版 2005 年 8 月第一次印刷
印数：1—10000 册
定价：**29.00 元**
ISBN 7-112-07560-2
(13514)

版权所有 翻印必究
如有印装质量问题，可寄本社退换
（邮政编码 100037）
本社网址：http://www.china-abp.com.cn
网上书店：http://www.china-building.com.cn

为什么
世界著名的商贸大厦都选择
SLOAN 仕龙

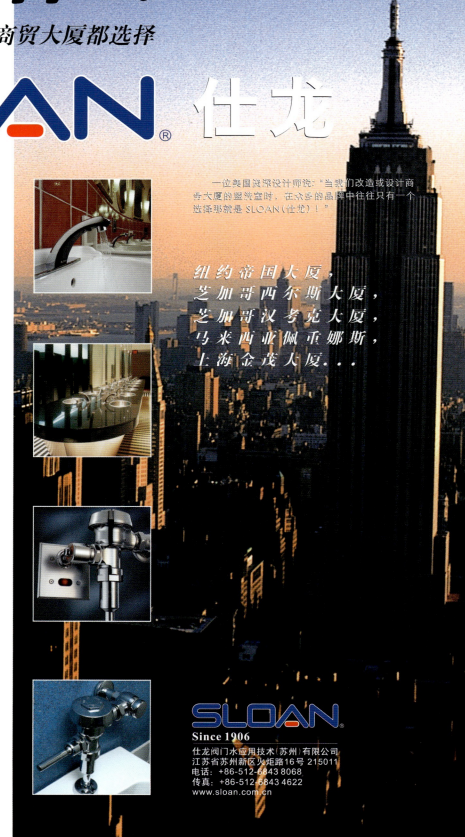

在美国，SLOAN（仕龙）的产品广泛应用于公共场所的盥洗室，包括商务大厦，运动场馆，娱乐中心，饭店，宾馆，商场，大学和医院。

"就业主而言，质量是首要考虑的因素。他们需要高质量、低维护的水龙头及冲洗阀来强化、改进卫生间的功能，与他们最高品质的商业大厦相匹配。当今，质量和性能是我们选择产品的关健。

选择SLOAN（仕龙），是因为其产品具有经典流体力学的可靠与现代电子技术的先进。SLOAN（仕龙），久经考验，富有创新和远见。"

所有的SLOAN（仕龙）冲洗阀，性能突出，具有超强的冲洗能力：

- 专利技术的双重滤膜装置及定量旁通孔的膜片适合不同的水质环境
- CID™ 技术确保定量冲洗，节约用水
- 真空止回阀确保污水不因水压下降而造成回流
- 可自我调节感应范围的红外感应技术及指示灯

当您面临高档卫浴、商务套房的设计困境时，SLOAN（仕龙）的电子感应龙头往往能带给您崭新的创意：

- 红外线感应技术保证公共卫生和健康
- 世界首创，最先进的光纤感应设计
- 可调温的电子智能模块一体化设计

既然许多世界著名的商贸大厦都**选择SLOAN（仕龙），那您为什么不呢？** 今天就打电话吧！

如果您想知道为什么世界最好的建筑都选择SLOAN（仕龙），请致电：+86-512-68438068 或访问SLOAN仕龙（中国）的网站：
www.sloan.com.cn.

一位美国资深设计师说："当我们改造或设计商务大厦的盥洗室时，在众多的品牌中往往只有一个选择那就是SLOAN（仕龙）！"

纽约帝国大厦，
芝加哥西尔斯大厦，
芝加哥汉考克大厦，
马来西亚佩重娜斯，
上海金茂大厦...

SLOAN
Since 1906

仕龙阀门水应用技术（苏州）有限公司
江苏省苏州新区火炬路16号 215011
电话：+86-512-6843 8068
传真：+86-512-6843 4622
www.sloan.com.cn

ARCHITECTURAL RECORD

建筑实录 年鉴 VOL.2/2005

封面：66号餐厅
设计：Richard Meier & Partners
摄影：Scott Frances/Esto
右图：环带混合住宅
设计：Steven Holl Architects

专栏 Departments

- 9   篇首语 Introduction
  激活城市公共生活
  by Clifford A.Pearson and 赵晨
- 10   新闻 News
- 70   翻译 Translations

专题报道 Features

- 14   中国住宅新策略 New Strategies for Housing in China
  三家国外建筑设计公司的高密度住宅项目创新设计
  by Clifford A.Pearson

作品介绍 Projects

- 22   W饭店，纽约 W Hotel, New York City
  Yabu Pushelberg 在世界上最繁忙的地方之一的时代广场营造出一片宁静
  by William Weathersby, Jr.
- 28   普赖斯大厦旅馆，俄克拉何马州 The Inn at Price Tower, Oklahoma
  Wendy Evans Joseph 的精品旅馆为赖特的摩天楼改成的艺术中心锦上添花
  by David Dillon
- 36   "Q!"旅馆，柏林 Q! Hotel, Berlin
  Graft 在柏林时尚街区的旅馆用包裹的表面活泼地演绎其浓厚的设计意味
  by Philip Jodidio
- 44   Megu，纽约 Megu, New York City
  Yasumichi Morita 在纽约翠贝卡区的时髦餐馆彰显日本文化的戏剧色彩
  by Clifford A.Pearson
- 50   帕提娜餐厅，洛杉矶 Patina, Los Angeles
  Belzberg Architects 在盖里的迪斯尼音乐厅内开创极具个性的餐饮空间
  by Joseph Giovannini
- 56   66号餐厅，纽约 66, New York City
  Richard Meier 在商业中心为名厨打造顶级餐馆赋予中餐全然现代的诠释
  by Suzanne Stephens
- 60   科略斯特里餐馆，瑞士格施塔德 Chlösterli, Gstaad Switzerland
  Patrick Jouin 让阿尔卑斯300年历史的小木屋变成时尚餐厅与迪斯科舞厅
  by Philip Jodidio
- 64   利华大厦餐馆，纽约 Lever House, New York City
  Marc Newson 让现代主义地标建筑的地下室变成50年代风格的时髦餐厅
  by Cynthia Davidson

1. W饭店，纽约；2. 普赖斯大厦旅馆，俄克拉何马州；3."Q!"旅馆，柏林；4. 帕提娜餐厅，洛杉矶；5. Megu，纽约；6. 66号餐厅，纽约；7. 科略斯特里餐馆，瑞士；8. 利华大厦，纽约

您可以在以下网站找到这些文章：www.architecturalrecord.com 或者 www.construction.com

敬请期待！
2006 全球建筑峰会
The 2006 Global Construction Summit

2006 年 4 月 25~27 日
中国，北京

2004全球建筑峰会吸引了超过450位世界各地行业精英齐聚北京。 麦格劳-希尔建筑信息公司与中国对外承包工程商会将再度联手打造2006全球建筑峰会。群英聚首，交流切磋，共拓商机！2006全球建筑峰会将是全球建筑及设计界领军人物的必到之盛会！

请立即报名，切莫错失良机！

大会报名，请登陆峰会网站 www.construction.com/event/

询问议程及参与发言， 请洽许敏达女士 minda_xu@mcgraw-hill.com

咨询赞助事宜， 请洽江大维先生 dave_johnson@mcgraw-hill.com

支持单位：

中国商业部
中国建设部
北京市政府

主办单位：

McGraw_Hill CONSTRUCTION

中国对外承包工程商会
CHINA INTERNATIONAL CONTRACTORS ASSOCIATION

connecting people_projects_products

McGraw_Hill CONSTRUCTION

在线咨询 www.construction.com

激活城市公共生活
Enlivening the Social Fabric of Cities

By Clifford A. Pearson and 赵晨

世界级的都市
需要吃喝玩乐的
休闲空间
也需要
便利宜人的
居住环境

最近的一次北京之旅中，我和当地的建筑师朋友们共进晚餐。餐馆坐落于后海一条古老的胡同内，曾经是四合院，现在成了人们餐饮聚会的休闲场所。尽管建筑本身并无特别之处，但是老建筑在这里找到了新的生机，正在成为这个国家首府的一幅充满活力的娱乐图景。在满足新用途的同时，它保留了老北京的尺度与城市肌理。对于像我这样的一位局外人来说，它让我窥见了北京城里正在飞逝的部分。

本期杂志介绍的餐厅与旅馆，看起来与这个四合院餐馆毫无关联，然而它们都有一个共同的特点，即在发现其价值的过程中吸引公众的注意。尽管餐厅与旅馆通常为私人所有，但它们在当今都市的社会生活中扮演着尤为重要的角色。它们通过赋予旧有建筑新的商业功能，把步行活动引向老街，从而激活日渐衰落的街区。它们是实验性设计的练兵场，建筑师被鼓励尝试新材料，用创造性的手段处理空间、照明以及平面构成，它们也是社会生活的聚集地，各色人等在此相聚相融。

在欧美国家，从事直接生产的人越来越少，越来越多的人转向了服务型工业，例如餐厅和旅馆。今天的餐厅旅馆不仅仅是人们用餐就寝的所在，也是娱乐与社交活动的聚集地。因此这类建筑必须能够聚集人气，就像剧院那样给人们提供视觉愉悦。从正门到吧台再到就餐空间的行进过程，仿佛是一种空间叙事；而别致的材料、戏剧性的色彩与照明，更能带来意外的惊喜。

飞速的发展改变着北京这样的大都市。在创造时髦而富有活力的吃喝玩乐场所的同时，它也提出了一个重要挑战，那就是如何安置新的外来人口，如何解决城市发展建设中办公、商业与居住用地的置换造成的搬迁问题。本期的专题报道记叙了三个富有挑战性的中国居住项目案例。

改变有时也带来冲突。那天饭后，我和东道主朋友们走向近旁的酒吧。与我们用餐的那家餐馆一样，这家酒吧也是由胡同里的老宅改建的，周围的四合院仍有市民居住。在入口处我们被一群居民堵住了。他们对酒吧的噪声深感恼火，再不肯接受深夜狂欢者的到来。

如何调和变化与持续、新与旧，是当今中国乃至全世界的建筑师、规划师们共同面临的关键任务。

斯蒂芬·霍尔在中国北京的
"环带混合住宅
（Linked Hybrid Housing）"中
创造了一个融合公园、
商店和餐饮的
富有活力的公众生活领域

新闻 News

计算机技术激发下的北京奥林匹克信息网络中心设计

北京奥林匹克公园的标志性建筑之一的"数字北京(Digital Beijing)",由一家在北京、深圳两地设有分公司的建筑师事务所——都市实践(Urbanus)设计。"数字北京"将成为区域性通信网络中心及北京2008奥运会的数据指挥中心。它毗邻赫尔佐格＆德穆隆(Herzog & de Meuron)设计的国家体育场和澳大利亚柏涛(PTW)设计的国家游泳中心,将于2005年底开工,预计在2007年底完工。"水立方"游泳中心与"鸟巢"国家体育场目前均在施工中。

除了作为奥运会的指挥机构,"数字北京"也是一座数字时代的博物馆、展览馆和办公楼。它是第一个委托给中国设计公司的奥运会标志性建筑,也是第一个在一场全球7家设计团队参加的竞赛中由中国设计公司中标的项目。

一群在纽约的中国建筑师刘晓都、孟岩、王辉、朱锫于1999年创立了"都市实践",然后回到中国。他们早期的项目大都在深圳,包括倍受好评的深圳规划与国土资源局办公楼。"数字北京"将在都市实践的名下建成,因此虽然朱锫后来退出并已在北京单独成立了自己的事务所,该项目从方案到施工的发展过程中仍将由朱锫负责调整。

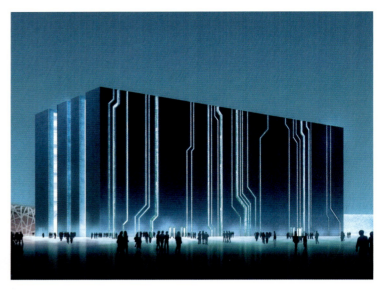

建筑的形式像计算机的主板(左图),立面则像硬盘(上图)。

建筑师们希望该项目建成一个数字时代的标志物,就像沃尔特·格罗皮乌斯(Walter Gropius)1911年设计的法古斯制鞋厂(Faguswerk shoe factory)标志了工业时代的到来一样,"如果工业革命对现代主义产生了巨大影响",朱锫问道,"那么数字时代会发生什么?"

"数字北京"力图通过缔造一个后工业时代的建筑来回答这个问题。在信息时代的审美意识影响下,它的立面与形状影射了计算机主板。建筑四个主要部分的特色在于效仿了计算机硬盘。首层平面用一个由许多桥划分的十字形的开放公共空间联系四个部分,并饰以电子影像。西立面使人想起条形码——一个在我们这个时代随处可见的符号,面向奥林匹克公园的立面则以一个巨大的传感器式电子显示屏投射色彩纷呈的图像来吸引人们的注意。

"都市实践"主持设计师王辉说,"它应该显示出人们正处于数字时代"。

"都市实践"现在正进行着其他各类项目的设计,包括北京和广东的一些商业和居住建筑,不过,"数字北京"是该事务所到目前为止最高调的项目。朱锫则在忙于北京北部的一个与张永和、齐欣等一批中国建筑师合作的软件园项目以及紫禁城边上的一个精品小旅馆的设计。

(D·艾尔斯 著,蔡瑜 译)

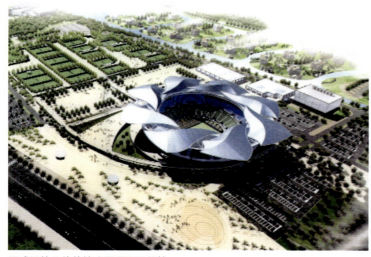

网球场的八片伸缩式屋顶可以旋转。

一座新的网球中心在上海绽放

位于上海市郊的旗忠森林体育场(Qi Zhong Stadium)一期是新落成的一座耗资2亿美元的网球中心,将于2005年9月开放。到2006年该项目的二期竣工后,旗忠网球中心将比目前亚太地区最大的网球场馆——墨尔本公园网球中心(Melbourne Park Tennis Center)的规模还要大。

主赛场可容纳1.5万人,高约40m,拥有一个独特的可开启式顶棚,由八片顶构成,形似上海市市花白玉兰。顶棚开启的时候,八片屋面缓缓旋开,从上空俯视,宛如一朵中心对称的八瓣花。

项目已设计完成并开始施工,由包括现代集团、中建三局、江南重工、上海机施在内的中国公司团队联合设计建造。上海机施负责建造主赛场的那个特别设计的屋顶。

这座位于上海市西南部、距上海市中心约27km、占地32hm^{2}的网球中心,将从2005年11月拉开帷幕,连续主办三年网球大师杯赛(Tennis Masters Cup)。除了主赛场之外,该中心还拥有40片室内、室外场地,包括一个规模稍小些的8000座的体育场。

(Jen Lin-Liu 著,蔡瑜 译)

新的总体规划与新建筑为汕头大学带来新气象

广东汕头大学邀请了多位外国建筑师进行校园设计,使其从传统整体的中式学院转变为更具学术交流氛围的学府,从而鼓励学生发挥更多的自由思考和独立自主的能力。

学校改造的中心部分是一座由台湾建筑师陈瑞宪(Ray Chen)新设计的图书馆,工程于今夏动工,预计于2006年底建成。3层高的图书馆形似一本直

立的半开的"书","书"的两部分间用坡道联系。该图书馆将容纳一个可供超过500名学生使用的自修室,还设有一个室内花园,陈预计,这样做可以使学生"不确定他们事实上是在室内"。

瑞士赫尔佐格&德穆隆(Herzog & de Meuron)建筑师事务所设计了校园总体规划,该方案给学生提供了自由活动的草坪、葱郁的大树和蜿蜒的小道,力图营造出"中心公园般(Central Park-like)"的宜人环境。[赫尔佐格的评论]

校园改造是汕头大学2002年引入教学改革以来最大的成果。虽然是一所公立大学,但是汕大还是得到香港慈善机构李嘉诚基金会的大力资助。赞助人香港著名实业家李嘉诚先生出生于汕头,并在1981年创办汕头大学。

据基金会高级项目经理梁先生(Fabian Leung)估算,基金会将在今后两年多为项目建设投资约3亿人民币,涉及范围从诸如健身房改造之类的小规模建设到图书馆等大项目的建设。梁说:"我们非常关注学习的环境,大学教师是这个环境中最直接的部分,但建筑是形成这个革新的基础。"

香港建筑师Matthias Woo和

由台湾建筑师陈瑞宪设计的新的图书馆计划于2006年底竣工(左图和上图)。

Wallace Cheung已经将现存的三幢建筑改造成名为"789"的学生活动中心,中心内设一个网吧咖啡厅、一个书店,还有乐队练习室以及给其他学生社团的小一些的空间。Woo和Cheung的设计使这三个建筑产生了两个庭院,庭院中穿插了盖顶的坡道,使这一综合体成为一个整体。Woo也是李嘉诚基金会的校园规划与建筑设计顾问,Woo说,"从原来的校园中央规划处变成新的学生活动中心,象征了校园中正发生着改变。"他认为,"这显示出校园是如何变得分散"以及开始更多地关注学生。

另一些校园中已建成的项目,包括会堂、宾馆及一组高级教师别墅,均由

张叔平(William Cheung)设计,于2002年完工。张叔平是指导《花样年华》的香港著名电影美术指导。学校2002年在教师别墅边上重修了一个蓄水池,水池由一系列台阶叠成,形成一个露天剧场,学生可以在此休息和学习。

汕头大学正在与日本建筑师长谷川逸子(Itsuko Hasegawa)及东京建筑事务所KDA(Klein Dytham Architecture)讨论一些校匠建筑单体的设计。计划到2007年建成一座新的宾馆用以接待重要来宾,还有两座新的教学大楼用以容纳新闻与传播学院和商学院。

(Jen Lin-Liu 著,董晓霞 译,蔡瑜 校)

澳门赌场新的主题策略——向拉斯韦加斯学习

拉斯韦加斯的扩张是如此之广,它甚至不远万里来到了中国。拉斯韦加斯金沙集团(Las Vegas Sands)最近宣布他们将在澳门兴建一处赌博娱乐度假区,它几乎就是闻名遐尔的拉斯韦加斯大道(Las Vegas Strip)上的威尼斯人赌场的翻版。

澳门的这座面积达100万m²、拥有3000间客房的酒店及赌场,同它在拉斯韦加斯的原版一样由达拉斯当地的HKS公司和威尔逊建筑师事务所(Wilson and Associates)共同设计,同样以模仿威尼斯的标志性建筑为特色,如总督府(the Doges Palace)和利亚德桥(the Rialto Bridge)。但是澳门度假区的规模大约是其拉斯韦加斯原版的三倍,正如HKS的主要负责人杰夫·詹森(Jeff Jensen)所说,"像一个发福的威尼斯人"。

该度假区将包括超过450m长的零售街(相当于拉斯韦加斯威尼斯人赌场中零售区总数的3倍)、超过4.5万m²的赌博空间(超过拉斯韦加斯赌场的5倍),以及一个1.5万座的用于展览和表演的剧场。

詹森的公司同时也在进行拉斯韦加斯三座赌场的设计,他说他的委托人在其第一个澳门项目的赌场中更倾向于使用一种成熟的主题策略,而非尝试某种更前卫的方式。"为什么要重新设计轮盘呢?我们就用已经成功运营的方案更好地与之配合吧。"

很多来自拉斯韦加斯的开发商希望通过在路水大道(Cotai strip)开发一系列的豪华娱乐场和酒店来提升澳门博彩区的特色。SOM(Skidmore, Owings, and Merrill)和易道(EDAW)为该地区设计的总体规划和建筑正在兴建[见《建筑实录》,2003年10月,第38

澳门的威尼斯娱乐场看上去与拉斯韦加斯的非常相似,不过更大些。

页]。赌博市场看来相当大。据詹森说,其委托人的前一个设在澳门的赌场——金沙娱乐场(The Sands)开业时,客人们蜂拥而至,几乎踏破门槛。

(S·卢贝尔 著,段巍 译,蔡瑜 校)

新闻 News

新建博物馆与改建老街区——唤回上海犹太遗存的历史记忆

上海市中心一片跨越7个街区的地区,在二战期间曾经居住了4万犹太人,如今将作为一个旅游区和综合社区重新开发。

上海市政府于2005年6月核准了由多伦多Kirkor建筑师事务所的克利福德·科曼(Clifford Korman)与中国同济大学共同设计的提篮桥地区总体规划的最终方案。加拿大活桥公司(The Living Bridge Corporation)于1月被获准独家开发该地块。

总平面中最引人注目的是Kirkor设计的一座犹太人博物馆,该建筑由四个立在水池上的立方体组成,它们之间用座玻璃桥连接。该博物馆将展览一些文件,诸如该地区建于1927年的犹太教会堂里举行成人仪式(bar mitzvahs)的邀请函,以及该地区1949年之前的照片。

这个海外投资者斥资7亿美元的项目要求有3座高层和几座小高层建筑用于商业、零售和居住,并由不同的建筑师来设计。依照科尔曼的规划设计,将使其中一条街禁止车辆通行而变成步行街,并保留或改建25%的现存建筑作为商铺和住宅。"我们的目的是加强两种不同文化之间的联系,加深对中国犹太人历史的了解",活桥公司的主席兼执行总裁伊恩·利文撒尔(Ian Leventhal)说。利文撒尔是一位专业的商业艺术家,他2001年来上海旅游时发现了这个地区的特殊价值。

2002年,利文撒尔和活桥公司主席托马斯·M·拉多(Thomas M. Rado)共同举办的上海犹太区艺术展,是活桥公司和上海市政府之间的首次合作。利文撒尔和拉多组建的活桥公司专门对北外滩地区进行了改造,该地区有维多利亚风格的联排别墅、一座装饰艺术派风格的剧院,以及上海独有的混合东西方传统建筑元素的石库门住宅。犹太人最早迁入该地区是在19世纪,主要来自巴格达。接下来是为了逃避迫害的俄国和波兰犹太人。到上世纪30年代的纳粹时期,第三次犹太移民潮中有高达2万欧洲犹太人涌入上海。

该项目将于2006年春动工,首期包括现存建筑的改建和犹太人博物馆的新建。整个项目预计赶在2010上海世博会之前完工。超过1.5万居民将要搬迁,其中大多数是老年人。
(Jen Lin-Liu 著,段巍 译,蔡瑜 校)

博物馆(上图)将适合改建后的街区(右图)。

北京新的特色社区——充满阳光的开放社区

巴黎建筑大师包赞巴克(Christian de Portzamparc)近日规划完成了地处北京东南部占地50hm²的社区。他还将继续深化设计一期工程的两个街区。该社区将提供8万m²的居住公寓和商业空间。整个社区被称为北京的后勤港口,目的是使其成为一个多功能的城市经济重点复苏区。这个区域将吸引许多高新产业及其从业者来此发展。

由亚洲最大的开发商之一——嘉里建设有限公司(Kerry Properties)开发的项目将成为首都规模最大的城市改造成果中的一例,并为城市生活提供一种新的模式。包赞巴克谈到,"这个项目的关键概念是'开放街区'",这一概念是他一直关注并从20世纪70年代以来在法国应用于不同项目中的一种设计理念。

该项目的核心是公私共有公寓(condominium)的集群,即在中国形成一种相对新的房屋所有权形式。包赞巴克提倡将一系列透气的、开放的街区穿插于整个社区之中,以创造一种比北京典型街区更符合人体尺度的空间。包赞巴克认为,"开放街区在一个高密度的居住项目中为创造符合人体尺度的空间提供了一种良好的途径"。

建筑师还设计了一种综合遮阳系统,使得建筑在整个白天都处在阴影之中,并以此吸引居民到室外活动。包赞巴克和他的团队设计的建筑单体形态各异、错落有致,使阳光能够渗入每个街区。将地块一分为二的中间绿带约略影射了中国北方的传统院落。在街道层面,将设置各式商业和服务空间,以强化社区内的活动,并使活动多样化。

该项目一期由包赞巴克主持设计,目前正处于当地政府的审批之中,并准备于2005年下半年破土动工。
(D·艾尔斯 著,董晓霞 译,蔡瑜 校)

包赞巴克将北京后勤港设计成一个综合社区,让室外空间渗透进每个街区。

Architectural, High Performance Outdoor Luminaires
建筑性 高性能的户外全配照明系统

Elegance
高雅

Excellence
杰出

Sophistication
先进

Character
个性

Distinction
卓越

KIM LIGHTING
16555 East Gale Avenue
City of Industry, CA 91745
www.kimlighting.com

Represented by:
International Lighting Asia (Hong Kong)
852-2310-8908

 Hubbell Lighting, Inc.

Made in America
美国制造

中国住宅新策略
New Strategies for Housing in China

By Clifford A.Pearson 蒋妙菲 译 蔡瑜 校

在过去的10年里，中国有超过1亿人口从农村迁入城市，形成了有史以来最大的城市化热潮。当前中国的城市人口正以每年2.5%的速度增长，赶超了世界上最快的增长率。人口统计学家预测，2050年城市人口将由2005年的3.5亿激增至8亿。与此同时，新富起来的城市居民想要提高居住的品质，一大批境外商业人士也渴望购房。从而使得大量移居的农村人口和成功人士的住宅供给成为亟待解决的问题。

从1998年开始，中国政府利用税收法和其他相关政策激励市民个人购房，并放宽了对外国人购房的政策限制。于是，高涨需求下的私营房产开发热潮开始涌现。在多数开发模仿不经济的美国郊区化居住模式的同时，也有一些更具改革精神的房产商在市区投资建设了高密度的住宅群，并请建筑师为之设计创新的居住理念。本文对此选取了目前正在进行的3个项目，其中两个由美国的事务所设计，一个由荷兰的事务所担纲，他们正致力于在城市的特定场所建造新住宅。

三家国外建筑设计公司的高密度住宅项目创新设计

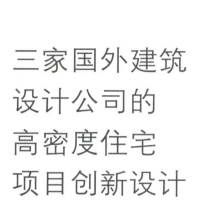

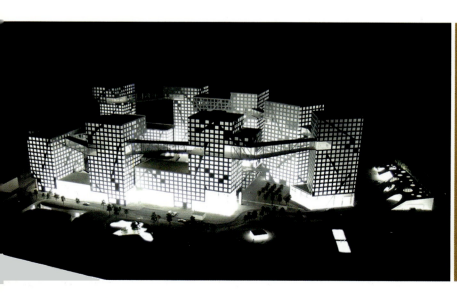

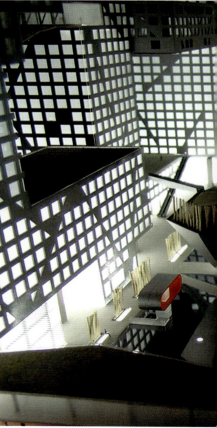

联合体环绕着一个的公园和一大片水池，但是提供了许多从外部进入社区公共空间的入口（左图）。旅馆设置在池中的一座圆形塔楼内（右下图），通过廊桥在上层与公寓相连（右上图、对页右上图）。

环带混合住宅，北京
斯蒂文·霍尔建筑师事务所

Linked Hybrid, Beijing
Steven Holl Architects

项目计划：680套公寓、12层旅馆、1000车位地下停车库、电影院、幼儿园、商店、餐馆、画廊、体育馆
业主：当代集团
时间：2005年9月—2007年9月

与许多房产商在中国城市开发独立高层住宅的做法相反，斯蒂文·霍尔与项目建筑师李虎设计了一个22层的高层住宅网，并连接地面公园和位于20层的一圈商店、画廊、咖啡馆及健身俱乐部。设计师没有把建筑群从所在的东直门街区中隔离出来，而是让它对近邻开放，并让多样的建筑与室外空间对话——这便是霍尔"渗透建筑"的概念。"这不是修道院"，霍尔说道。对于方案鼓励邻里共享花园与公共空间的做法，他认为"渗透"概念具有社会价值和形态彰显的重要意义。

建筑师们还引入创新的"绿色"技术，譬如打钻660个100m深的地热井来供应总计5000kW的能量以采暖制冷。再生水循环系统将供给320m³的冲厕用水以及景观绿化与种植屋面的灌溉用水。

建筑的混凝土外壳既是围护又是结构，取代了梁柱系统，为内部空间的自由划分提供了最大的灵活性。混凝土外壳应用绝缘技术后，用铝板包裹出建筑的表皮。所有的窗套腹面以及建筑之间的连桥与坡道的外表都采用源自佛教建筑的色彩，装饰得生动活泼。每栋住宅楼都是一梯四户，户型面积从75m²至300m²不等，内部空间的灵活性还使得开发商可以提供约50种不同的房型。

空间的动线设计对于人的感受和体验至关重要，霍尔说："我们把这个空间序列设想成电影场景，从连桥到公园，公园到公共空间，再到公共空间彼此之间，都会获得一系列景观视野。"来访者可以从3个主入口进入公共空间：被大片倒影水面环绕的电影院、圆形的高层酒店，以及其中的一座住宅塔楼。房产商最初设想建筑只是单纯的住宅，但建筑师认为，加入影院和画廊等文化功能以及强化室内外的共享空间极其重要。当代集团的总裁张雷同意扩展项目，为创建愉悦的整体城市和共享生活空间而积极开发。"他年轻、聪颖，而且有些理想主义。"霍尔这样评价他的业主。

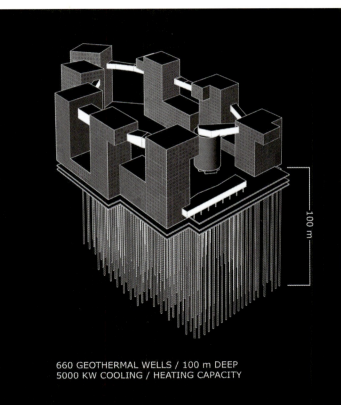

660 GEOTHERMAL WELLS / 100 m DEEP
5000 KW COOLING / HEATING CAPACITY

项目引入创新性的"绿色"技术，如使用地热井在夏季将冷气送入建筑，冬季送暖气（上图）。窗套腹面以及建筑之间的连桥都使用了缤纷的色彩（左图、左上图、右图）。每栋住宅楼都是一梯四户，户型面积从75m²至300m²不等。户型安排使得每一户都有一个两面开窗的拐角，能拥有两个不同方向的视角（右上图）。

西钓鱼台塔楼住宅，北京
伯纳德·屈米建筑师事务所
West Diaoyutai Tower, Beijing
Bernard Tschumi Architects

项目计划：180套公寓、365套客房的旅馆、健身俱乐部、花园、餐馆、商店、游艇俱乐部
业主：N+N房地产投资公司
时间：2006年春－2007年秋

有别于周围常规的住宅建筑，伯纳德·屈米将这个5.9万m²的项目设计为在顶部悬臂搭接的双塔，赋予其引人注目的识别性。这两个塔楼——一座是公寓，一座是酒店——在接近地面处倾斜分开，形成多功能的裙房，里面容纳了商店、餐厅、健身俱乐部、游艇俱乐部和花园。建筑师们进一步对群房的设计提出两种设想：一是"生态外皮"，3层的裙房连同顶部都进行绿化；二是立面隐喻，采用印满花叶图案的玻璃幕墙诠释对绿色的渴望。虽然第二种方案看起来少一些创意，但考虑到城市空气的污染程度，这个方案更具可行性。两种方案都会将底层处理成花园，并在室内外运用水体优化环境。

建筑师们采用打满三种不同口径圆孔的钢板给建筑设计了精致的表皮。摒弃了谨慎保守的立面，屈米设计的建筑外观恰似一个均匀布满图案的巨大封套，将不同楼层和居住单元隐藏并笼罩起来。对此屈米解释说："这就好比中国传统的屏风或围墙，用连续的几何图案营造出围合感，却并不阻碍采光和通风。"建筑坐落于昆玉河畔，毗邻玉渊潭公园，赏获水景正是近水楼台。而通过金属面板和玻璃的包裹，他们也想赋予建筑鳞光闪闪般的外表。

屈米与布罗·哈普达（Buro Happold）工程公司合作设计了建筑的钢结构以及顶部控制整体结构的巨大桁架。建筑共有27层，其中包括3层的裙房和2层的地下室。

在塔楼里设置旅馆，是希望创造一个有活力的社区，这里的北京居民可以同游客和外宾一起度过愉快的时光。公寓单元包含80至95m²的单卧室户和100至120m²双卧室户两种房型，所有的公寓都共享酒店的家务管理和门房等服务。酒店除提供260套标间之外，还有80间套房和25套复式房。酒店与公寓拥有各自的独立入口，一个拥有圆形天窗、3层通高的中庭为住户与旅客创造出动人的共享空间。

屈米认为，高层住宅在北京和纽约通常都由开发商来决定居住模式，因而往往是大同小异，"我们试图打破惯例，向高层住宅的思维定势提出挑战。"

龙潭公园，柳州
MVRDV 建筑师事务所

Longtan Park, Liuzhou
MVRDV, Architects

项目计划：2700 户居住单元
业主：柳州 He Jia 房地产公司
时间：2005 年底 —（待定）

基地位于柳州市郊一个石灰岩采石场，五座美丽的喀斯特山体由于矿石的开采被一截为二。荷兰的MVRDV建筑事务所决定用新建筑使遭到破坏的环境重获新生。MVRDV的策略是用联排复式住宅将悬崖断壁层层包裹，在抑制土层侵蚀恶化的同时造就景观与建筑的完美合一。设计还保证了每户都拥有山谷美景和自然通风。

联排住宅与石崖之间留有3m的空隙，设置楼梯和环廊，并保证空气流通。在不同层高上设计了平台，按照自然地形的走势确定住宅的梯度。建筑师针对不同的地质特征提出不同的建造方法：硬岩的峭壁上房子要用销钉加固；较为平缓的硬岩上建筑要用柱子支撑；而岩质疏松和劈裂的地方是不能建房子的，所以就设计了广场和平台，作为户外公共空间。

建筑师计划用混凝土板建造，在混凝土和墙体中掺入当地的矿石，从而使建筑与其环抱的山崖融为一体。建造中使用的销钉、五金构件以及建筑间隙处的平台，在整个建筑复合体中创造出"一种迷人的空间差异"。所有住宅都是复式房型，由上层或下层进入，从而减少了走廊和电梯平台的数量。

在山顶将种植树木花草，并蔓延至联排房屋户外空间的每个角落，有效地保护了建筑与环境。MVRDV敏锐地在遭到破坏的环境中插入新建筑，力图以建筑来抚愈大自然受到的人为创伤。

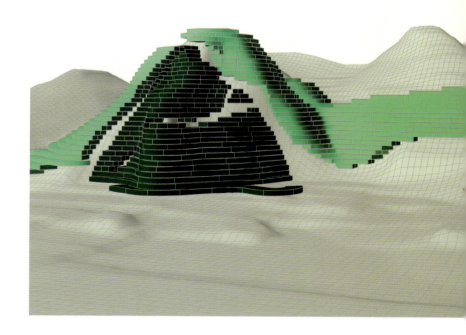

建筑师希望通过台地住宅修复因为采矿而受损的山体（上图、左图）。所有住宅都是200m²左右的复式房型。建筑依山就势（上上图），处于悬崖陡坡和或疏松或坚硬的石阶上（右图、右上图）。

Cabanalike enclosures rim the lobby lounge, blurring the line between public and private (above). Mirrored partitions and red-painted canvas backdrops frame a lobby platform (below) and the check-in stations (opposite).

At the **W HOTEL** in Times Square, designer *Yabu Pushelberg* creates spare, richly textured compartments, evoking a bento box's sensual delights

By William Weathersby, Jr.

It's a long way from Planet Hollywood to the stylish sphere of the W Hotel in Times Square. The 52-story tower was initially planned as a frenetic, family-oriented satellite of Planet Hollywood's theme restaurant chain. After a change in ownership during construction, however, the pop-culture setting was eclipsed by a more refined realm. With interiors reinvented by the interior design firm Yabu Pushelberg, the W Hotel's serene landscape almost floats above the streetside chaos at the so-called Crossroads of the World.

Replacing a mass-appeal menu, the hotel's interiors now suit subtler tastes. Like a lacquered bento box, the Zenlike hotel offers an array of delights imbued with natural color and texture, all contained in interlocking enclosures. It's a sampler of varied moods, functions, and materials—spare vignettes forming a harmonious whole.

When Planet Hollywood filed for Chapter 11 in 1999, Starwood Hotels & Resorts Worldwide stepped in to retrofit the building as a flagship for its W brand, a domestic chain of 16 urban hotels geared to design-savvy travelers. Starwood scrapped the existing interior specifications and commissioned the Toronto team of George Yabu and Glenn Pushelberg to rework public spaces and guest rooms, providing, as Pushelberg puts it, "an oasis from the sensory overload of the theater district."

The designers tackled the building's constraints head-on. With tight floor plates, boxy public areas were stacked on the lower floors with few connective or transitional spaces. At street level, separate entrances to the lobby, restaurant, and below-grade nightclub further broke continuity. Complicating matters, the restaurant and nightclub were leased to outside operators, introducing yet another client tier to review the design.

"The hotel's public spaces fit together like pieces of a puzzle," Pushelberg says. Patrons might visit only one area, perhaps the restaurant or one of four bars, while a registered guest might require a range of experiences to enrich a weeklong stay. Working with the preset infrastructure, Pushelberg explains, "we explored the idea of compartmentalization, guiding guests through a series of varied, unfolding spaces."

Though the project's populist Hollywood iconography had been abandoned, "we were aware of Times Square's 'everyman' connotation at the heart of the city's busiest tourist district," Pushelberg says. Employing

Contributing editor William Weathersby, Jr., frequently writes about lighting and interior design for RECORD. He lives in Westport, Conn.

Project: W New York Times Square, New York City
Owner/client: Starwood Hotels & Resorts Worldwide
Architect of record: Brennan Beer Gorman Architects—Mario LaGuardia, AIA; Kevin Brown
Interior designer: Yabu Pushelberg—George Yabu, Glenn Pushelberg, principals; Mary Mark, Reg Andrade, Anson Lee, Marcia MacDonald, Cherie Stinson, Aldington Coombs, Alex Edward, Eduardo Figuerero, Marc Gaudet, Mika Nishikaze, Sunny Leurg, Kevin Storey, project team
Lighting designer: L'Observatoire—Hervé Descotte, principal

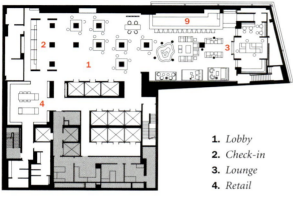

1. Lobby
2. Check-in
3. Lounge
4. Retail

LOBBY: SEVENTH FLOOR

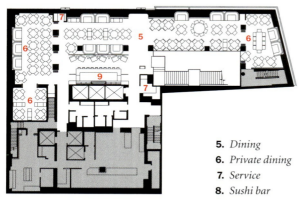

5. Dining
6. Private dining
7. Service
8. Sushi bar

BLUE FIN RESTAURANT: SECOND FLOOR

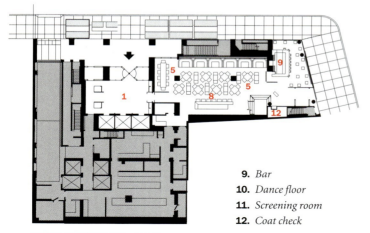

9. Bar
10. Dance floor
11. Screening room
12. Coat check

BLUE FIN RESTAURANT: GROUND FLOOR

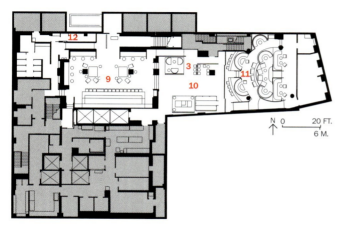

WHISKEY BAR: CELLAR LEVEL

In the Blue Fin restaurant, pine-slat partitions create intimate settings (left and opposite, top). Enhancing the terrazzo staircase, a textured concrete "Wave Wall" and a mobile of abstract fish subtly support the aquatic theme (below left and opposite, bottom). Illuminated jigsaw sections of mirror above the banquettes echo the compartmentalized plan of the two-floor space (below right).

Chain-link curtains undulate through The Whiskey, a below-grade disco-styled nightclub (right and opposite, top). Dubbed the Living Room, the hotel lobby displays a range of upholstered furniture illuminated by oversize lanterns and a back- and front-lit bar (below and opposite, bottom). White-stained wood flooring unifies diverse seating areas in the lobby's loftlike space.

spare, linear furnishings and enclosures without gimmicks, "it's meant to be an accessible, nonconfrontational Modernism that welcomes everyone."

As with Yabu Pushelberg's recent redesigns of interiors at Tiffany and Bergdorf Goodman stores in New York, the W juxtaposes a range of striking visual "moments." Entering the elevator lobby on 47th Street, guests cross a dramatic threshold. Framing the foyer, water flows behind glass panels along the ceiling and side walls, immersing the compact space in refracted light and the sound of trickling water. A black terrazzo floor and steel elevator doors sharpen the cool edges.

Reaching the seventh-floor reception lobby, elevators open directly onto a lounge called the Living Room. Oversize light boxes hover above white leather-upholstered ottomans marching through the loftlike space. The monochromatic palette—white-stained oak floors, terrazzo platforms, acrylic cubes, and plaster and resin walls— is accented by colored lighting cast across an expansive bar and by a rainbow-hued painting in a niche behind an acrylic jigsaw screen. Since the existing windows would have overlooked an adjacent office building, the designers obscured them for a calming, cocoonlike ambience. Seating options— from rough-hewn wood stools on the periphery to deep, plush sofas in an "inner sanctum" near the bar—invite different modes of interaction.

To the west of the living room, the registration area is set within a darker rectilinear "container" defined by black terrazzo flooring, dark paneling, and ribbed-mirror-glass insets. Behind the compartmentalized counter, a slash of red-painted canvas gives the walls a varnished look. Tucked around the corner, a gift shop appears as a brightly lit white cube.

Fronting Broadway and accommodating two cocktail bars plus a sushi bar, the two-floor Blue Fin restaurant belies its 400-seat size. Slats of pine baked to a blackened finish are strung to form sculptural partitions that screen more intimate vignettes. Mirrored walls acid-washed in amber, charcoal, and red tones softly reflect light on bare wood tables circled by Saarinen chairs. Along the terrazzo staircase, a textured concrete wall of undulating waves ascends to a mobile of an abstract school of fish floating overhead. The aquatic theme is sketched in subtle strokes.

In The Whiskey, the 6,000-square-foot basement-level nightclub, chain-link curtains cloak seating areas and a separate screening room, outfitted with mod leather-covered banquettes and stools. A vintage '70s dance floor, whose liquid-filled panels change color as dancers step across them, is an homage to a disco that once occupied the site.

Upstairs, each of the 509 guest rooms is spare yet sensual. Furnishings in smoky charcoal and tobacco shades are set against muted wall covering with color gradations from slate to sand. A continuous neutral plane—a composite of resin and terrazzo—flows from the entry floor into the bathroom, and up along the walls to wrap the shower. Translucent polycarbonate screens, cutting across a corner of the bathroom, serve as room dividers, blurring public/private boundaries.

Like Yabu Pushelberg's playful spaces throughout the W Times Square, these rooms have turned the standard decor of Midtown chain hotels inside out, packing a variety of surprises into a stripped-down, refitted container. ■

Sources
Wall covering, wall finishes: *Metro Wallcoverings; Moss & Lam; Excelsor*
Cabinetry, millwork: *Benchmark Furniture; Pancor Industries; Erik Cabinets*
Paints, stains: *Sherwin Williams*
Flooring: *Stone Tile International; Sullivan Source*

Lighting: *Abramczyk Studio; Baldinger; Color Kinetics; Eurolite; Sistemalux; TPL Marketing; Unit Five Manufacturing*
Furniture: *Knoll; Minima; Pancor*

www For more information on the people and products involved in this project, go to Projects at architecturalrecord.com.

Wendy Evans Joseph turns an iconic work by Frank Lloyd Wright into THE INN AT PRICE TOWER with no edginess lost

By David Dillon

On the 9th of February in 1956, Bartlesville, Oklahoma, residents lined up for blocks to tour Frank Lloyd Wright's bizarre addition to their downtown. Soaring sunscreens, thrusting balconies, no right angles: how to make sense of it? Locals referred to it as a spaceship, a hood ornament, and "Price's folly," a reference to Harold C. Price, who had given the commission to Wright.

"People thought it was crazy because it didn't have walls until the very end," recalls Price's daughter-in-law Carolyn. "It looked like the backbone of a fish. But over the years they've come to like it. You could say it went from crazy eccentric to wonderfully eccentric."

Bartians (rhymes with Martians) are hoping that the old magic continues now that the tower has been converted to an arts center, with a 21-room boutique hotel designed by New York architect Wendy Evans Joseph, AIA. The town is still reeling from the recent departure of the Conoco/Phillips headquarters and in need of a boost. The Inn at Price Tower could be it. Where else can you visit a Frank Lloyd Wright skyscraper and also get to spend the night?

Harold Price had made a fortune in the pipeline business and wanted to give something back to his hometown, something special, a new civic icon perhaps [RECORD, February 1956, page 153]. So in 1952, Wright proposed a pinwheeling, 19-story, poured-concrete tower, with floors cantilevered from a central core like branches on a tree and supporting a combination of offices, apartments, and shops. It was an elaboration of his unbuilt St. Mark's in the Bowerie project of 1929, though with an asymmetrical plan and more complex program. Taliesin apprentice Fay Jones contributed to the working drawings.

"Tap root" schemes turn up repeatedly in Wright's work, including this exotic, mixed-use project in a remote Oklahoma oil town. The Bartlesville tower is an architectural summa of Wright's views on integrated living and working, the liberating power of technology, and the primacy of instinct and emotion. It is as much a social manifesto as a work of architecture.

Harold Price balked at first. He had been thinking of a two-, maybe three-story build-

Contributing editor David Dillon is architecture critic of The Dallas Morning News.

Project: *The Inn at Price Tower, Bartlesville, Oklahoma*
Design architect: *Wendy Evans Joseph Architecture—Wendy Evans Joseph, AIA, principal; Robert Furno, AIA, Farzana Gandhi, Manan Shah, Thruston Pettus, Liz Burrow, Liza Beaulier, design team*
Client: *Price Tower Arts Center—C.J. Silas, chairman; Richard P. Townsend, executive director*
Architect of record: *Ambler Architects—Scott Ambler, principal; Jim Charles, engineering technician/project manager*
Furniture, guest-room rugs, and mural design: *Wendy Evans Joseph Architecture*

PHOTOGRAPHY: © CHRISTIAN KORAB

In 1956, Frank Lloyd Wright's 19-story Price Tower in Bartlesville, Oklahoma, was the tallest thing around. Seen at first as an alien object, it soon became a familiar icon, and now the top floors have been turned into a boutique hotel.

1. Entry
2. Lobby
3. Museum reception
4. Museum theater
5. Temporary exhibition
6. Hotel reception
7. Service entry
8. Guest room
9. Guest suite, lower level
10. Kitchen
11. Conference room
12. Restaurant, lower level

TYPICAL GUEST FLOOR

FIFTEENTH FLOOR

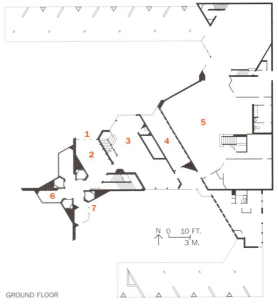

GROUND FLOOR

ing, he told Wright. "Too much wasted space," the architect replied." "But it's such a big building for a small town," Price continued. "Not at all," said Wright. "I just stood it on end."

And so it went, with Wright ultimately getting everything he wanted, even a drive-up branch of the Public Service Company of Oklahoma. (The Art Center's main gallery sits where the pay window used to be.) He also busted Price's budget by $1 million.

But dismay turned to delight. Price put the tower on the cover of *Tie-In*, the company magazine, and rarely passed up a chance to tout its brilliance. Employees reportedly liked the building. "You weren't stuck in a cubicle," remembers Bill Creel, head of the company's pipeline-coating division. "You could get up and walk around and look at the sky and the prairie. You get a lot of good ideas with views like that." But others weren't so impressed. The duplex accommodations on one quadrant of each floor turned out to be small, hot, and expensive, too much like Manhattan apartments. Price later converted most of them to offices, which proved unworkable because of so many triangles and so little flexibility.

The Price Company occupied the tower until 1981, then sold it to Phillips Petroleum,

IN WRIGHT'S MIXED-USE TOWER, THE APARTMENTS TURNED OUT TO BE TOO SMALL, HOT, AND EXPENSIVE. SO THEY WERE CONVERTED TO OFFICES.

Originally, each level of the poured-in-place-concrete structure was a combination of single-level offices and duplex apartments. In this incarnation, Wendy Joseph converted the top eight floors into a 21-room hotel, with three duplex suites (above left). The color scheme, materials, and furniture that Joseph designed (opposite) defer to Wright's own aesthetic, while creating a new identity for the inn.

Visitors to the hotel and arts center can take an elevator from the restored lobby (above) to Copper, the bar and restaurant (left three and opposite) on the 15th and 16th floors. In remodeling the interior, Joseph designed its furnishings and murals to diverge from the triangular form Wright used, in order to keep the two approaches distinct. The curved bar itself (left) has a copper sheet top and is faced with maple plywood and Lucite strips. The chairs and bar stools are maple and copper pipe, while the banquettes are maple.

which used it for 20 years before turning it over to the Price Tower Arts Center, a nonprofit institution for contemporary art and architecture. Founded in 2001, the center has an ambitious agenda of exhibitions and public programs, which it hopes to fund in part with income from its hotel and restaurant. It is also trying to raise $15 million for a new museum by Zaha Hadid [RECORD, June 2003, page 30], a low horizontal building that will wrap around the tower like a boomerang while reaching out to a performing arts center designed by Wright's son-in-law, William Wesley Peters. Perhaps Hadid's bold geometry will spark another public debate and the familiar/unfamiliar cycle will begin all over again.

Wendy Evans Joseph's hotel carries on a spirited dialogue with Wright's building without lapsing into either mimicry or glibness. She complements, at times contests, but never copies. It wasn't easy, she explains, because "the building is so specific, with such strong geometry and so few materials, that I wondered if there could be another response. I wanted to leave Wright alone, yet I found I had to contend with him at every turn."

She approached this strictly as an interiors job. She made no structural changes to the tower; the spaces remain as Wright designed them, but with different contents. Eighteen offices and three of the original apartments have become guest rooms ($125–$250 per night), with a fourth turned into the restaurant Copper. The lobby, complete with Wright's favorite quotation from Walt Whitman, has been faithfully restored, along with Harold Price's office on the top floor, which looks as if it had been shrink-wrapped the day he departed.

Joseph succeeds in preserving the spirit of Wright while updating the materials and technology. She uses abundant copper in the guest rooms and restaurant, for example, but in the form of tubing and thin industrial mesh instead of heavy patinated panels. Likewise, her carpets and upholstery fabric recall Wright motifs without copying any particular one. His metaphor for Price Tower, "the tree that escaped the crowded forest," served as the starting point for the contemporary, even slightly Japanese, murals in the guest rooms.

Wright's tables and chairs were exercises in pure form, miniature sculptures, while hers consist of long thin strips of maple that highlight construction and invite rearrangement. His seem anchored; hers seem to float in their spaces. Because the tower's elevators are the size of phone booths, the individual pieces had to be hauled up one by one and assembled on-site, the same approach Wright had used.

WENDY JOSEPH CARRIES ON A SPIRITED DIALOGUE WITH WRIGHT'S BUILDING WITHOUT LAPSING INTO EITHER MIMICRY OR GLIBNESS.

The visual high point is Copper, a two-level restaurant on the 15th and 16th floors that was formerly an apartment. The bar is a long, gentle curve with lapped maple sides—a wry allusion to Wright's swirling Guggenheim—and a faceted copper top. The dining tables are covered with a thin copper mesh slipped between sheets of glass; the same mesh is used for the window drapes, providing a fluid counterpoint to the heavy copper sunscreens on the outside of the building. The room is like a three-dimensional sculpture in which no two surfaces, and no two views, are the same.

With the completion of this $2.1 million renovation, the Price Tower has come full circle—from curiosity to avant-garde symbol to corporate castoff to community icon. Its prototype, St. Marks, was all apartments; its most recent incarnation is a residence of a trendier and more transient kind. "It is a unique opportunity to honor the spirit of Wright," says Arts Center director Richard Townsend. "It's also a chance to create the kind of community focal point, the town square, that Harold Price had in mind." And, in the process, to reintroduce the public to one of the most romantic, idiosyncratic, and fiercely individualistic works of architecture in America. ∎

Sources
Fabric seating: *Maharam*
Fabric for draperies: *Jack Lenor Larsen*
Guest-room lamp lighting: *Lite-Source*

For more information about this project, go to Projects at *www.architecturalrecord.com*.

The nonprofit Price Tower Arts Center occupies the first two floors of Wright's only skyscraper. Here, Joseph created an architecture and design gallery (above) that features Wright's objects. The original 19th-floor office for Harold Price, with the furniture designed by Wright, has been kept intact as part of the museum (opposite).

With undulant, folded planes, Graft animates the interior of Q!, a new hotel in Berlin, sensuously flowing walls into ceilings and furniture

By Philip Jodidio

Look for luxury goods shops in any self-respecting European capital, and the immediate corollary, the "design hotel," can't be far away. Sure enough, in Berlin, around the corner from Chanel, Louis Vuitton, and Cartier, a new hotel so chic its entrance carries no name proves the rule. But for those who know where to look, Q! isn't hard to find. Within a relatively undistinguished gray facade, punched windows (of the sort commonly seen in this city) reveal the first hint that Q! stands apart from its neighbors in this high-rent district, just off the Kufürstendamm. From the street, white translucent curtains veil the hotel desk, but just through the glass front doors, the visitor enters a sea of wraparound, red surfaces that look more like California than New Berlin.

Flowing from the lobby into the lounge and restaurant, curvy, red-linoleum-clad surfaces glide seamlessly from floor to wall. Couches and built-in furniture similarly bear the mark of the architects, Graft, a young Los Angeles- and Berlin-based firm with a total of 20 employees. Its partners, Wolfgang Putz, Thomas Willemeit, and Lars Krückeberg, owe this commission more to Hollywood connections than German origins. Hotel operator Wolfgang Loock called them in 2002 after seeing an article on the Hollywood Hills studio they'd designed for actor Brad Pitt.

By the time Graft took on the interiors for this hotel—the firm's first—its developer had already selected an architect for the shell of this seven-story building, though construction had not yet begun. Graft shared little of that architect's sensibility, but managed to work with him nonetheless. After proposing a different facade (which was never realized), the partners accepted the delicate task of executing a challenging project with a low budget (approximately 1 million euros for the interior)

Philip Jodidio is a Paris-based journalist and the author of more than 20 books on contemporary architecture.

Project: Q!, Berlin
Architect: *Graft*—Lars Krückeberg, Wolfram Putz, Thomas Willemeit, partners; Wolfgang Grenz, project leader; Johannes Jakubeit, Michael Rapp, Sasha Ganske, project team; Stephanie Bünau, Sven Fuchs, Lennart Wiechell, Leo Kocan, Nikolas Krause, Helge Lezius, participants

Visible from the exterior (opposite), folded red planes enliven the interiors. Inside, floors meld into walls, ceilings, and furniture, as at the bar (right).

Undulant red-linoleum-clad planes bend to form high and low perches, illuminated recesses (right and opposite, bottom), and the front desk (opposite, top). Graft also designed the fireplace and couches (below).

The spa, in the basement, includes a sandy "beach" lounge/cinema (left and bottom left) and black-terracotta-lined showers (opposite). Here, sculptural though rectilinear walls catch light and shadow, while incorporating shelves (opposite and below right).

1. Reception
2. Lobby/lounge
3. Restaurant
4. Bar
5. Buffet
6. Kitchen
7. Standard guest room
8. Front entrance

TYPICAL UPPER LEVEL

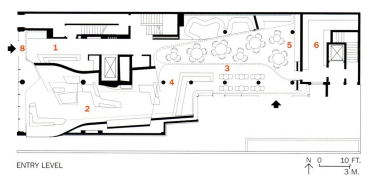

ENTRY LEVEL

and a client perhaps not initially attuned to their ideas. Further constraints included an elevator and stair core already in place near the entrance and a small footprint, measuring just over 4,000 square feet.

Throughout the 32,000-square-foot building, Graft succeeded in imposing a unified aesthetic—one admittedly influenced by the folded planes of architect Neil Denari, who became director of Southern California Institute of Architecture (SCI-Arc) just after Putz and Krückeberg had completed their studies there. At Q!, continuous, streamlined surfaces wrap not only the street-level public areas but also the guest-room interiors, where the palette shifts to white against smoked-oak floors. Here, walls meld into desks and ceilings. Overhead, curved ceilings lightly printed with Christian Thomas's photographs of a woman, aim to give these quarters what Putz calls a "cocoonlike feeling."

Carefully thought out, the room designs favor an enveloping smoothness that does away with many everyday clues to designated function, such as door handles. Cupboards or light switches are not immediately visible. According to Putz, "We want the visitors to take a moment to orient themselves." Though sleek wrappers seem to be de rigueur in new design hotels, Graft crafted the aesthetic skillfully, conveying a sense of high quality through good workmanship, despite the low budget. What looks like slate in the bathrooms, for example, is really black terra-cotta, and so forth.

While the pale ceiling photos may recall Jean Nouvel's more forceful integration of movie images into his design hotel in Lucerne, Switzerland [RECORD, May 2001, page, 238], the Graft architects claim to have found inspiration elsewhere. They liken their work to a film storyboard—cinematic in its aspirations, whereas most Berlin architecture, suggests Putz, tends more toward still photography. The experience of working with Brad Pitt, he says, influenced Graft to consider architecture in these narrative terms. Whether in the bar or the guest rooms—where the bathtub sometimes melds with the bed—the architects imagined the space as a movie director might, envisioning the scenes with guests moving through the interiors, or sets.

Q! attempts to bring to Berlin the kind of style and design-consciousness of Philippe Starck's Saint Martins Lane (SML) hotel in

London [RECORD, January 2000, page 90], or other Ian Schrager properties, without spending a fortune. Although the Graft partners say they've never seen the inside of SML, it appears that the influences of Starck and Nouvel, in addition to Denari, have somehow filtered into Q!. Both the hotel's design and its service give a distinct impression of déjà-vu. But perhaps what seems refreshingly Californian in Graft's approach is its spirit of openness and optimism, transcending what the partners call "typical German skepticism." Instead of rejecting this project as impossible, with its tight space and means, the architects flowed the smooth curves of contemporary design into a hard-angled, gray Berlin box—no small feat. ■

Sources
Linoleum: *Marmoleum*
Furniture: *Vitra; Fussgestell; Alias; Moroso; Paola Lenti; ArtifortLande; La Palma; Tischplatte*
Sinks: *Duravit*

Shower fixtures: *Dornbracht*
Tiles: *Atala; Sicis*

For more information on this project, go to Projects at
www.architecturalrecord.com.

In many guest rooms, a bed and tub merge into a single, curvy whole, daringly juxtaposing activities that conventionally occupy separate spaces (this page and opposite, top and bottom).

Megu
New York City

YASUMICHI MORITA BRINGS HIS HIGH-ENERGY BRAND OF MODERN JAPANESE DESIGN TO AMERICA AND GIVES A SHOWSTOPPING PERFORMANCE.

By Clifford A. Pearson

Architect: *Kajima Associates*
Interior designer: *Glamorous Company—Yasumichi Morita, Satomi Hatanaka, Seiji Sakagami, project team*
Owner: *Koji Imai/Food Scope New York*
Engineers: *Hage Engineering (structural); CY Mills (m/e/p)*
Design consultant: *Hashimoto & Partners—Osamu Hashimoto, Sachiko M. Masaki, project team*
Consultants: *Kenji Ito (lighting); Shoji Tahara, SKS Scott Kirk/Carlo Fornerino (acoustical)*
Construction supervisor: *Toshi Enterprise*
General contractor: *Kudos Construction*

Size: 14,000 square feet
Completion date: *March 2004*

Sources
Cabinetwork and woodwork: *Cmack Construction*
Wall and floor tiles: *Seto Seikei*
Chairs: *Lef*
Vinyl leather upholstery: *Sincol*

For more information on this project, go to Projects at
www.architecturalrecord.com.

When Megu opened in Tribeca this March, it made a big splash on the New York restaurant scene. The food, the service, the design, and the prices are all larger-than-life, as if made for the silver screen. Rocco DiSpirito's one-year-old restaurant on 22nd Street might be reality TV, but Megu is a Technicolor fantasy.

If your idea of Japanese restaurants was shaped by the blond woods and graceful counters of small sushi bars, Megu will come as a shock. There's nothing quiet about this place, from the waitstaff yelling "*Irasshaimase!*" (welcome) as you arrive in the dining room to the bold colors and unorthodox mixing of materials all over the two-story establishment. Call it Modern Japanese Baroque. The design certainly matches the food, which includes such showy dishes as Kobe beef cooked at the table on sizzling hot rocks and salmon-and-toro tartare with a mound of wasabi-soy mousse that's melted in front of your eyes by a waiter holding a red-hot iron poker.

The man behind Megu is Koji Imai, a 35-year-old entrepreneur who has 30 restaurants in Japan. With Megu, his first foray into the American market, Imai hopes to kick-start a run of restaurants in New York and perhaps other parts of the U.S. To lead the design team for his American flagship, Imai hired

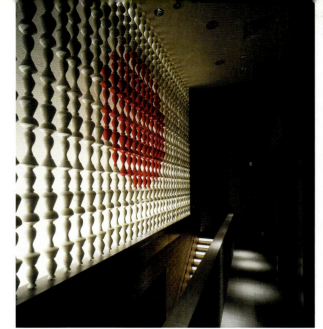

Rising sun: A Japanese flag made of porcelain sake bottles stacked on rice bowls grabs attention from the street (opposite, left) and separates the entry foyer from the stair leading down to the dining room (right). At the landing, a grid of sake labels works as art (far right). In the bar, the designer used kimono fabric in rolls on two walls and spread out on the overhead light (below).

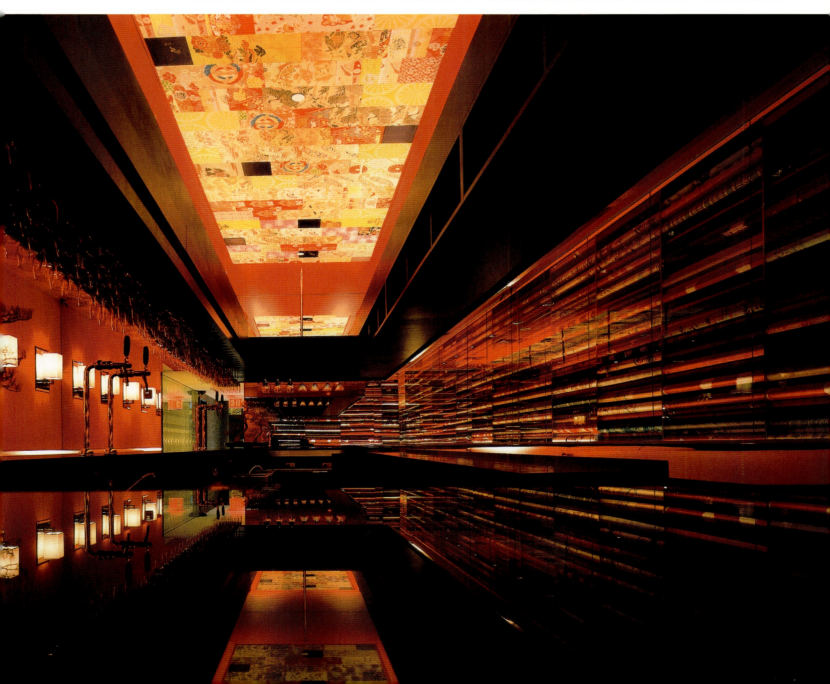

Yasumichi Morita, a young Osaka-based designer who had worked with him on Maimon, a restaurant that opened in Tokyo's Shinjuku district in 2002.

Program

Part of a new generation of supersize restaurants opening in Manhattan, Megu sprawls over 14,000 square feet and includes a vermilion-colored "Kimono Bar," an "Imperial Lounge" overlooking the dining room, a small VIP lounge originally conceived as a smoking room, a sushi bar, and a private dining room adjacent to the kitchen, in addition to the 200-seat main dining room. The restaurant occupies the ground floor of a 19th-century cast-iron building and flows into the basement level as well.

Solution

"Because Megu is so big, we designed it as a series of different scenes," explains Morita. The action begins on the sidewalk, where guests can see a backlit, mosaiclike wall in the foyer emblazoned with a red Japanese sun in the center. Closer inspection reveals the wall to be made of porcelain sake bottles and rice bowls stacked one atop the other so they form columns. Like the first shot of a well-crafted movie, the entry wall provides important clues about what comes next. Reinterpreting icons of Japanese culture and using old materials in strikingly new ways turn out to be key themes tying together Megu's conspicuous displays of imagination.

After the porcelain bottle-and-bowl wall, the first full dramatic scene happens in the bar, where rolls of kimono fabric line two walls, and squares of the same fabric form a kind of quilt stretched over a long light box above the bartenders. Morita used mirrors and the room's vibrant Chinese red to crank up the impact of the luxurious kimono material, creating a dazzling, almost kaleidoscopic effect even before customers order their drinks.

The designer skillfully alternated action scenes with quiet moments, such as the lounge just

You rang? A 700-pound bell copied from one at a temple in Nara, Japan, hangs from the ceiling of the 40-foot-high dining room.

PHOTOGRAPHY: © NACASA & PARTNERS

beyond the bar, where beige mercedes leather and tall curving banquettes set a relaxed tone. He also choreographed the experience of moving through the restaurant; for example, directing customers down a paired set of narrow stone stairs, so the double-height dining room looks even bigger when they arrive at their tables.

At almost every turn, Morita found yet another ingenious way of treating familiar materials. On the way to the restrooms, customers walk past a wall of Japanese matchbook covers set into glass. At the stair landing, they can admire a grid of sake labels attached to curving plastic mounts and lit from behind. In the dining room, the designer created a checkerboard of bamboo mats on one wall, and on the opposite side he glued thin rectangles of stone on glass so they seem to float in an old masonry pattern.

Holding center stage in the dining room is a giant, 700-pound bell, a facsimile of a much heavier one at a temple in Nara, Japan. Sitting below is a Buddha ice sculpture, slowly melting into a pool decorated with floating hibiscus leaves. Bordering on kitsch, the bell and Buddha serve as a visual anchor to the large dining hall.

Commentary
Moving beyond stereotyped images of geishas and samurai, Morita and his client have translated Japanese culture into an architectural language understood by New Yorkers. Tactile, bold, and inventive, their restaurant engages diners in a cinematic experience that unfolds as one moves from one space to another. "Megu is not just for eating," says Morita. "It is also entertainment." That it is.

At a time when people jet around the globe and images bounce instantly from one continent to another, Megu offers a high-energy interpretation of modern Japan by a Japanese artist for an American audience. Is it authentic? Does it make any difference? It's show biz. ∎

GROUND FLOOR

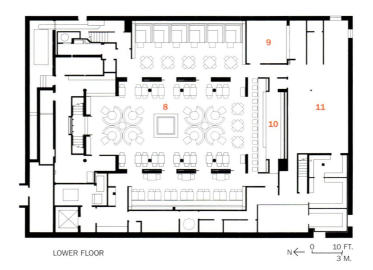

LOWER FLOOR

Never-ending cycle: Every day a new ice Buddha must be made (opposite, top). The sushi bar features a colorful image of Nara printed on glass (opposite, bottom left). For the west wall of the dining room, Morita glued stone on glass (opposite, bottom right). On the east, he created a warmer surface using bamboo mats (above).

1. *Entry hall*
2. *Reception*
3. *Coat check*
4. *Bar*
5. *Lounge*
6. *Pantry*
7. *VIP lounge*
8. *Dining*
9. *Private dining*
10. *Sushi bar*
11. *Kitchen*

Patina Restaurant
Walt Disney Concert Hall
Los Angeles

HAGY BELZBERG, AIA, BRINGS FOLDS OF WOOD AND CONTOURS OF GLASS TO A RESTAURANT AND CAFÉ IN L.A.'S DISNEY CONCERT HALL.

By Joseph Giovannini

Architect: *Belzberg Architects—Hagy Belzberg, AIA, principal; Eric Stimmel, Manish Desai, Erik Sollom, Ryan Thomas, Melanie Freeland, Dan Rentsch, Leyden Yaeger, Jaron Lubin, project team.*
Consultants: *John Dorius Associates (mechanical); A+F Consulting Engineers (electrical); Daniel Echeto (structural: restaurant); John A. Martin Associates (structural: building); Tom Nasrollahi and Associates (plumbing); Martin Newson Associates (acoustic); Michael Blackman Associates (kitchen); Elizabeth Paige Smith (colors and materials)*
Contractor: *Matt Construction*

Size: *5,000 square feet for 112 seats plus 48 on the terrace (restaurant); 4,000 square feet for 200 seats (café)*
Cost: *Withheld*
Completion date: *October 2003*

Sources
Wood millwork: *Mueller Custom Cabinetry and Spectrum Oak Products*
Carpet: *Bloomsburg Carpet*
Chairs: *Holly Hunt*
Tables: *West Coast Industries (dining room); Janus et Cie (café)*
Stainless-steel fittings: *CR Laurence*
Surfaces: *Dupont Corian*

For more information on this project, go to Projects at **www.architecturalrecord.com**.

The commission to design a restaurant and café in Walt Disney Concert Hall was fraught with potential pitfalls. The design could neither upstage Frank Gehry's masterwork nor play possum. The most treacherous misstep could turn this corner of Disney Hall, along Los Angeles's Grand Avenue, into ersatz Gehryland. "We wanted to be respectful, but we had to have our own identity," says Hagy Belzberg, the Santa Monica architect chosen from a long roster of local designers who submitted credentials for the coveted work.

Program
Belzberg, whose previous restaurants and houses have been materially rich and spatially robust, had to work within Disney Hall's prescribed shell. Here, he inherited a plate-glass facade: little more than a recessed strip subordinated to Gehry's streaming forms.
    Existing doorways, already positioned in the facade, became starting points for Belzberg's plan for the 5,000-square-foot ground-floor restaurant and 4,000-square-foot café.

Solution
As built, the restaurant's main entry opens onto a waiting area in front of a bar. Wine bottles, arrayed

Joseph Giovannini practices architecture in New York City, teaches at UCLA, and is New York magazine's architecture critic.

with their corks forward, form a pattern behind a backlit, translucent wall. A small private party room lies to one side of the entry. A translucent curtain veils the room's sidewalk views, while a picture window offers glimpses into the kitchen. To the other side of the entry, a large dining room appears in dark, muted, low-contrast colors.

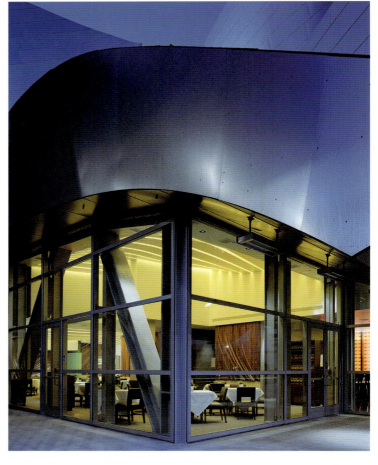

Sliding, square wall panels display art curated by the Los Angeles Museum of Contemporary Art across the street.
    For Belzberg, computer-enhanced metaphors proved the best defense against the mythic, potentially overwhelming presence of the Gehry building—a historic icon even before its inauguration

PHOTOGRAPHY: © TOM BONNER

Computer-milled, solid walnut panels in the dining room evoke rippling curtains. Bent luan plywood strips create an undulant ceiling that recalls an awning.

Channels between the billowing contours of bent plywood strips contain cold-cathode lighting (top and opposite), to which Belzberg added copper film to create a sunny glow in contrast to the dark, walnut wall panels (bottom).

last September. For Belzberg's high-end restaurant (with its tenancy undecided even through its construction process), the allusion to a theatrical curtain inspired the design. This venue, after all, would be most active before curtain time and after performances. The absence of a literal curtain in the performance hall—which has a thrust stage instead of a proscenium—gave the metaphor a provocative edge.

Curtains, however limp, come to life as they are drawn and released. And the notion of freezing such motion inspired Belzberg to play that effect against the implied motion, or sailing forms, of the Disney facades.

Belzberg also responded to Gehry's monumental building through the application of computers in the design process. Gehry had famously used the aeronautical program CATIA to evolve the physical models for Disney Hall that he had first developed by hand with construction paper. Belzberg chose to enlist the computer, with form-Z software, to generate rather than confirm shapes (see sidebar, page 104).

Visually evoking stretching motion, Belzberg created and froze ripples on-screen and transmitted them to milling machines, which carved rigid "curtains" from solid panels of walnut two-by-fours laminated side by side. He placed these curtains through much of the restaurant, defining open spaces through slipped asymmetrical configurations. The ripples play against the walnut's grain and the stock lumber's module. Translating billowy "sails" into wavy "fabric," the panels restate Gehry's larger gestures in miniature. Abstractly, Belzberg reduced to a decorative

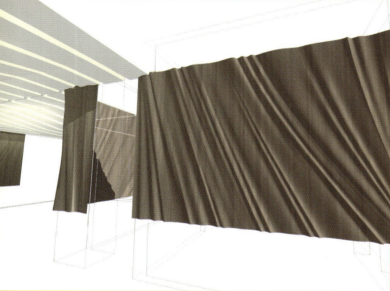

In addition to the dining room (plan, below), with 160 seats and a bar (opposite, bottom right), Belzberg designed a more casual venue, a café for 200. Sandblasted ½-inch-thick tempered glass surrounds its kitchen, creating a luminous volume (opposite). The café and restaurant are distinct in character.

scale the turbulent forms roiling through Gehry's building.

The ripples are deepest on the ceiling, where an undulant surface of curving luan plywood strips is illuminated by cold-cathode lights embedded in its interstitial troughs. A great wave of striated light, the ceiling provides a transitional scale between the panels and Gehry's majestic curves.

Commentary

Belzberg has created a well-behaved interior that is compatible with Disney Hall, relating to it formally, spatially, and technically—but without being acquiescent. Its computer-generated hybrid of curvilinear and orthogonal forms translates the building's macro gestures onto a microscale. Against Gehry's metaphors, Belzberg established his own. Yet the congruence between the two approaches allows for a seamless transition. Distinguished respectfully from Disney Hall, the restaurant design favors understated differentiation over obvious opposition. ∎

1. *Main dining room*
2. *Private dining*
3. *Restroom*
4. *Kitchen*
5. *Chefs' dining*
6. *Outdoor patio*
7. *Bar*
8. *Catering kitchen*

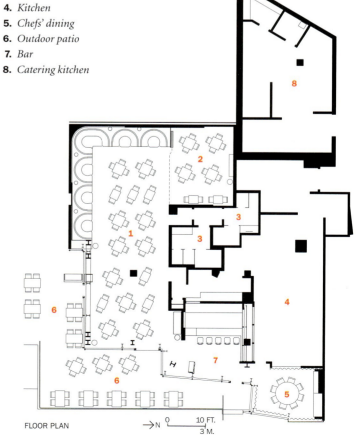

FLOOR PLAN

THE ONLY CURTAINS IN THE CONCERT HALL ARE MADE OF ... WALNUT

Go ahead, touch them: These wooden "curtains" are meant for your haptic pleasure. To make them, the designers created 3D models in form-Z software, then used the digital models to drive a CNC milling machine that carved them from 800-pound blocks of laminated walnut two-by-fours. Each block was 3 to 4 feet wide and 8 feet tall. The curtains were finished with a clear varnish and attached to the walls ("very carefully," notes design principal Belzberg drily) with steel angles and ledgers. The learning curve was steep for the firm's first foray into 3D design, Belzberg says, "but we had a patient and trusting client—and the fact that Frank Gehry approved the designs was a big help, too." He's enthusiastic about doing more work of this stripe. "As architects, we can now use software to sculpt spaces," he says. Eat your heart out, Claes Oldenburg.
Deborah Snoonian, P.E.

66
New York City

RICHARD MEIER STICKS TO WHITE (WITH TOUCHES OF RED) FOR A CHINESE RESTAURANT IN NEW YORK CITY'S TRIBECA NEIGHBORHOOD.

By Suzanne Stephens

Architect: *Richard Meier & Partners—Richard Meier, FAIA, principal; Don Cox, AIA, Thomas Juul-Hansen, design team*
Client: *Suarez Restaurant Group with Jean-Georges Enterprises*
Consultants: *Ambrosino DePinto & Schmieder (m/e/p); Goldstein Associates (structural); Mark Stech-Novak (restaurant consultation and design); L'Observatoire International (lighting)*

Size: *5,400 square feet, 150 seats (plus 25 in the lounge)*
Cost: *Withheld*
Completion date: *Spring 2003*

Sources
Glass partitions: *Architectural Glass Craft*
Stainless-steel mesh: *GKD-USA*
Absorptive acoustic-plaster ceiling: *RPG Diffusor Systems' BASWAphon*
Dining chairs: *Eames by Herman Miller*
Lounge chairs: *Cassina*
Stools: *Bertoia by Knoll*
Lounge tables: *Saarinen by Knoll*
Tables: *Atta Studios (custom tabletops); Gratz Enterprises (custom table bases)*

For more information on this project, go to Projects at
www.architecturalrecord.com.

Since 66 opened last spring, its Shanghai-chic cuisine by Jean-Georges Vongerichten and Minimalist modern interior by Richard Meier, FAIA, have garnered the fervent attention restaurateurs crave. In this case, the clients are both the chef, Vongerichten, and Phil Suarez, also one of the investors in the Richard Meier–designed Perry Street Apartment towers in the West Village. Even though Meier had only designed one restaurant before—for the Getty Center in Los Angeles, where competition from other restaurants on its hilltop site is not formidable—Suarez was not deterred. "We knew Meier would provide the right excitement for Vongerichten's cuisine," he says. (Nevertheless, Vongerichten's restaurant consultant, Mark Stech-Novak, was on hand to plan the kitchen and advise on other such matters, while the lighting consultant, L'Observatoire International, made sure the lighting would warm up Meier's renowned white palette.)

Program
Since Vongerichten and Suarez already operate a slew of restaurants uptown (Jean Georges, Vong, JoJo), the two decided on the ground floor of the Textile Building, a toned-down, Classical-style structure in Tribeca. Designed in 1901 by Henry Hardenbergh, the architect of the Dakota apartments and the Plaza Hotel, the landmarked building is not too far from Odeon, a pioneer of downtown arty-elegant restaurants, which opened almost 25 years ago. During this time, Tribeca has become a residential-loft paradise catering to the affluent who like the casual lifestyle with concierges.

Accordingly, Meier thought the restaurant should be open and light. "I wanted people to experience a degree of intimacy as part of a larger space," he says. "And," Meier adds, "I thought there should be no hierarchy in the dining spaces. Wherever you sit, you feel this is the most important spot."

Solution
Meier divided the rectangular space into three main sections around a central entrance vestibule, defined by a 12-foot-high, curved-frosted-glass wall. Floor-to-ceiling frosted-glass panels partition the various areas,

A curved, 12-foot-high glass wall (below) separates the communal table from the entrance vestibule (opposite). Behind the table, a frosted-glass wall (right) partially conceals drink preparation.

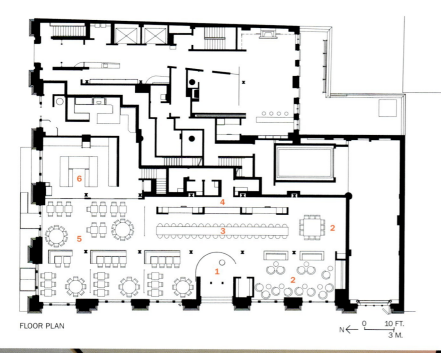

1. Entrance vestibule
2. Lounge
3. Communal table
4. Bar
5. Dining
6. Kitchen (final prep)

The 44-foot-long communal table (below) has an epoxy resin top and stainless-steel base with Bertoia stools. Chinese ideograms on red silk banners visually lower the 12-foot-high ceiling.

FLOOR PLAN

which are further subdivided by built-in stainless-steel-mesh cubicles with wood-panel and leather banquettes.

Behind the entrance vestibule, a 44-foot-long communal table seating 40 acts as the orienting locus in the restaurant, dramatized by a row of red silk banners hung from a slot in the dropped acoustic-plaster ceiling. The bar at the back of the communal table is concealed behind a frosted-glass wall, through which the bartenders' shadowy movements and the bottles' contours offer only ghostly traces of their presence.

In the dining area, the kitchen can be glimpsed through four glass water tanks containing vividly polychromatic fish. The immaculately organized kitchen is devoted mainly to the final stages of cooking; Halogen downlights prevent a harsh glare from being admitted to the dining room. (A second preparatory kitchen, for slicing and dicing, is located in the basement.)

Commentary

The combination of loft-renovation (with painted riveted-steel columns) and carefully designed dining alcoves shows the masterful attention to detail and craft for which Meier is known. The desired openness combined with intimacy is handsomely achieved through the translucency of the glass partitions and the gleaming stainless-steel-mesh cubicles, all of which subtly enhance the sense of elegance. The dominant use of white works well because of the softness of the ambient lighting (including candles at night) and splashes of color (red flags, variously colored fish).

Although the long communal table seems to be a fad appealing mainly to a cell-phone culture that thrives on strangers listening to private conversations, it seems to go over well (for now). The high culinary standards—supported by highish prices—have made 66 a magnet, even for those looking for a lunch spot while serving jury duty. Will it match the longevity of the less-expensive Odeon and its open and active bar scene? We'll have to see. ■

Four fish tanks divide the dining area from the kitchen (right). Sconces attached to the backs of wood and leather banquettes provide ambient lighting that flickers through the stainless-steel mesh of the squared U-shaped cubicles (below and far right).

Chlösterli
Gstaad, Switzerland

PATRICK JOUIN TURNS AN ALPINE CHALET INTO A CHIC DINING AND ENTERTAINMENT VENUE FOR EUROPE'S JET-SETTERS.
By Philip Jodidio

Designer: *Patrick Jouin—Patrick Jouin, Laurent Janvier, Tomoko Anyoji, Sanjit Manku, Tania Cohen*
Architect: *Robert Stutz*
Client: *Michel Pastor, Delphine Pastor*
Consultants: *Hervé Descottes (lighting); Philippe David (graphics)*
General contractor: *Michel and Delphine Pastor*

Size: *16,000 square feet, including two 1,100-square-foot dining areas, a 1,600-square-foot dining terrace, and an 850-square-foot discotheque*
Completion date: *December 2003*

Sources
Video fireplace: *Souvenirs from the Earth*
Wood terrace tables: *Michel Poupion*
Terrace chairs: *Fermob*
Armchairs in bar: *Cassina Contract*
Spoon chairs: *Cassina France*
Lighting: *SES Giraudon*
Stone paving: *Christian Messerli*
Wood flooring: *Müller-Hirschi*
Wine wall: *Chambrair*
Metal joinery: *Metalbau Stoller*

For more information on this project, go to Projects at
www.architecturalrecord.com.

Set in a 300-year-old chalet on the main road into the Swiss mountain resort of Gstaad, Chlösterli blends tradition, modernity, and a sense of humor. The chalet, built by the monks of Rougemont Abbey, had been converted into a restaurant and pizzeria before the Monaco developer Michel Pastor bought it. Pastor and the chef Alain Ducasse called on Paris designer Patrick Jouin to breathe new life into the dark wood structure. Jouin, who also worked with Ducasse on the Plaza Athenée Restaurant in Paris as well as Mix in New York City, is a 37-year-old who had been in charge of furniture and product design for Philippe Starck before starting his own firm in 1998.

Working within strict guidelines on what is the oldest wood building in the village, Jouin cleaned and restored the chalet's facades. The most visible intervention outside the building is a new, 1,600-square-foot terrace for summer dining made of Iroko wood and concrete. Subtle variations in the placement of slats in the wood enclosure surrounding the elevated terrace allow diners to take in the bucolic mountain setting.

Program
Ducasse's plan called for not one but two restaurants: a traditional Swiss dining venue on the ground floor and, above that, Spoon des

Philip Jodidio is a Paris-based journalist who writes about architecture.

Neiges, one of seven Spoon locations around the world. (Jouin designed the Spoon Byblos in Saint Tropez, which opened in 2002.) Ducasse also operates acclaimed restaurants in Paris, Monaco, and New York, and châteaux and hotels in France. Busy guy.

Each of the restaurants at Chlösterli has its own 2,250-square-foot kitchen serving a dining area of less than 1,100 square feet. Targeted to a wealthy clientele, Chlösterli includes an 850-square-foot discotheque on the ground floor.

Solution
Using the chalet's dark-wood interior as an aesthetic baseline, Jouin applied an unexpected mixture of modernity and tongue-in-cheek respect for Swiss tradition. Diners in the ground-floor restaurant see relatively little of the project's contemporary personality, entering the dining room from discreet, streetside doors and eating in a room where slate floors and oak paneling set the tone. Jouin reworked traditional oak chairs with saddlelike leather seats, gently tweaking convention. (Jouin designed all of the project's furniture and light fixtures.)

The two-story-high disco is the most spectacular departure from the usual Alpine experience. Scottish slate on the floor gives way to resin blocks lit from below by a LED system that pumps vibrant and changing colors into the space. A 17-foot-high glass wall divides the disco from the kitchen and serves as a giant, transparent wine rack. Jouin played on the incongruous presence

PHOTOGRAPHY: © THOMAS DUVAL

A new dining terrace (opposite) is the only major change to the exterior of the old chalet. Inside, LEDs light up the disco floor and a glass wall displays wines (this page).

of international sophistication in a traditional farming area by designing tables in the shape of old wine buckets and wood seats that are wry updates of vernacular prototypes.

Two cramped stairways, recalling the chalet's rural origins, take diners up to Spoon, where a sleek, Modern aesthetic asserts itself. In the bar, a "fireplace" made of plasma screens shows flickering images of the fire not allowed by local regulations. Metal-frame chairs slung with leather seats signal the more refined atmosphere on this floor, while a private dining area, nicknamed "the aquarium," offers views of the disco floor through a floor-to-ceiling glass wall. The second floor's entirely Modern vocabulary completes Jouin's sly transition from Switzerland's past to Gstaad's jet-setting present.

Commentary

Instead of denying or covering up the irony of a hip dining-and-partying venue in a house built by 18th-century monks, Jouin employed it as a design tool. Not wanting to erase the past but to play on it, he created a handsome and witty environment that takes diners on a spatial journey toward progressively more Modern settings and furnishings. Given the extremes involved, making this transition work without causing aesthetic gears to screech was no small task. Patrick Jouin pulls off the trick with cool panache, in the process bridging a gap of three centuries from timeworn wood to the pulsing beat of a discotheque. ∎

SECOND FLOOR

GROUND FLOOR

1. Bar
2. Discotheque
3. Entry
4. Traditional restaurant
5. Kitchen
6. Lounge
7. Spoon restaurant
8. Private dining
9. Office
10. Tunnel to terrace

Tables imitating wine buckets (above) and blocky wood chairs (far right and opposite, top right) in the traditional restaurant are sly references to rural prototypes. A private dining room (right) overlooking the disco floor and Spoon restaurant (opposite, bottom) features sleeker, more Modern furnishings. A banquette in the traditional restaurant (opposite, top left) offers a cozy place to relax.

Lever House Restaurant
New York City

MARC NEWSON INSERTS A STYLISH, FUTURISTIC FIFTIES RESTAURANT TO THE LANDMARK LEVER HOUSE.

By Cynthia Davidson

Interior Designer: *Groupe Marc Newson, Paris*—Marc Newson, principal; Sébastien Segers, consulting architect
Client: *Joshua Pickard, John McDonald, Robert Nagle, Aby Rosen*
Architect of record: *CAN Resources*—Taavo Somer, Derek Sanders, Serge Becker, Judy Wong, project team
Associate architects: *MGZ Architecture*
Consultants: *Vogel Taylor Engineers (mechanical engineering); L'Observatoire (lighting); Super Structures (structural)*

Size: *6,500 square feet, 130 seats*
Cost: *$5 million*
Completion date: *August 2003*

Sources
Custom seating and furniture: *Meritalia (Como) according to designs by Marc Newson*
Bar counter and tunnel surface: *Dupont Corian*
Carpet: *Durkan*
Granite: *Cold Springs Granite*

For more information on this project, go to Projects at **www.architecturalrecord.com**.

The value of the Lever House as a Modern icon on New York's Park Avenue was recognized when the city's Landmarks Preservation Commission designated the building a landmark in 1983, even though the Skidmore, Owings & Merrill design was only 31 years old (hardly an antique). Appropriately, by its 50th anniversary in 2002, the building was nearing complete restoration and rehabilitation [RECORD, March 2003, page 122], but bringing it back to life required more than new lobby furniture and curtain wall. A critical issue for lease-holder RFR was to animate the ground-floor space formerly occupied by a conference room and Lever Brothers company store.

Program
Enter New York restaurateurs John McDonald and Josh Pickard, who opened the Lever House Restaurant in August. The available 6,500-square-foot space is actually subterranean and windowless but accessible directly from 53rd Street on the south side of the building. The frontage available for establishing the restaurant's identity is minimal, and landmark laws prevent excessive signage on the building. Then designer Marc Newson came on board, an Australian (living in Paris) with a reputation for things curvilinear—bikes,

Cynthia Davidson is the editor of Log, *a new publication of observations on architecture in the city.*

chairs, airplane interiors, the "stuff that surrounds you"—with a retro Modern aesthetic that *Wallpaper* magazine has made so fashionable.

Solution
In less than three years, Newson concocted a pod of hexagons and curved surfaces that is both retro (fitting for a 1950s mothership) and very now. Working with in-house consulting architect Sébastien Segers, he created windows in the windowless space by lining one side of the room with large curved openings that resemble the windows in passenger trains. Diners step through them to sit at curving banquettes and look back at the crowd on the floor 6 inches—but feeling much farther—below. A large opening in the wall at the far end spans nearly the width of the room, framing a private, 22-seat dining room. This window is fitted with sliding sheets of clear glass that when closed provide acoustic, but not visual, privacy; hence diners here are always onstage, a twist on the

Visitors enter the main dining room (this page) after passing through a 20-foot-long tunnel lined in Corian. At the rear, an elevated dining area overlooks the space.

Rounded booths trimmed in blond oak line the west wall of the main dining room (above and opposite, top). Visitors enter from the street through a dark vestibule where the wine is displayed (right), before entering the white tunnel. A glass wine cabinet and a view of the main room enliven the 22-seat dining room (opposite, bottom two).

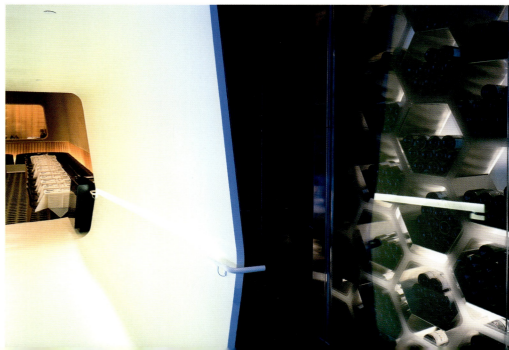

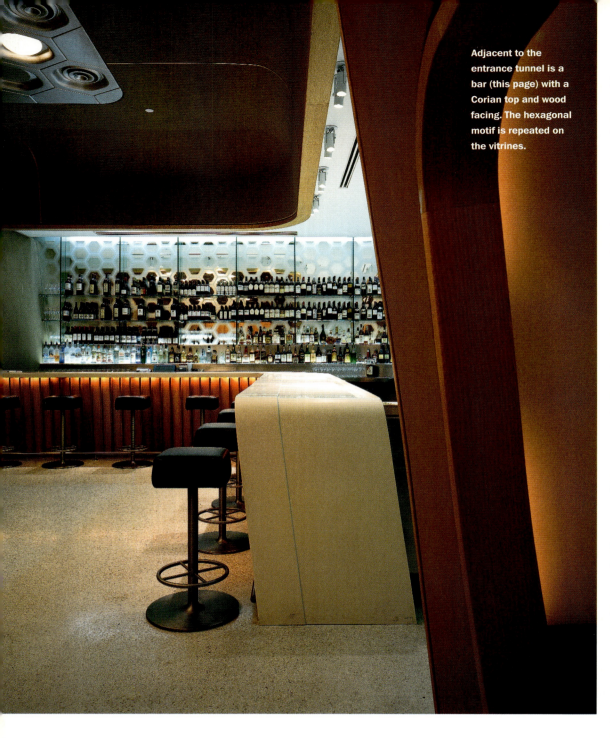

Adjacent to the entrance tunnel is a bar (this page) with a Corian top and wood facing. The hexagonal motif is repeated on the vitrines.

idea of dinner theater.

At the bar and in the dining room, Newson uses banal materials—white Corian; rough, putty-colored plaster; blond oak; mirror glass—with a high-style sensibility. The lightness of these materials and the curves Newson introduces to the room are highlighted by a completely black, orthogonal entry off 53rd Street, where coats are checked, and at the back, a completely black corridor leading to all-black restrooms (fixtures and all). The blackout look hides the damage that occurs with intensive use of the spaces, but more important, the darkness heightens one's sense of passage into the light, central space. The honeycomb of hexagons underfoot on the carpet, overhead in the coffered ceiling and private dining room lighting, as well as behind the bar, simply add geometric amusement for the eye.

Newson sets the scene for the action with an illuminated curved tunnel of white Corian that descends from the street-level lobby to the below-grade main dining room. This passage transports diners—like astronauts—into another dimension, where, unaware of eating in what is nearly a basement, the diners offer themselves to chef Dan Silverman.

Commentary

Lever House Restaurant opened in August 2003. On a Tuesday evening in September, the room was humming with an overflow crowd; without a reservation, one must wait 2 hours for a table on what is reportedly the slow night of the week. Is it the food or the ambience that the throngs are seeking?

The lobster tempura and roasted wild salmon are delicious, but no more impressive than the food served at the historic, Philip Johnson–designed Four Seasons Restaurant only one block away. That center for the power lunch seems to be the logical Lever House precedent. The difference in the decor is notable. Where Johnson's rooms continue to ooze a certain elegance, Newson's are more pop. Their lack of subtlety seems to destine them for a much shorter lifespan. ∎

1. *Entry*
2. *Coat check*
3. *Tunnel*
4. *Bar*
5. *Main dining room*
6. *Private dining room*
7. *Entrance to kitchen*
8. *Passage to Lever House lobby*
9. *Lever House exterior courtyard*

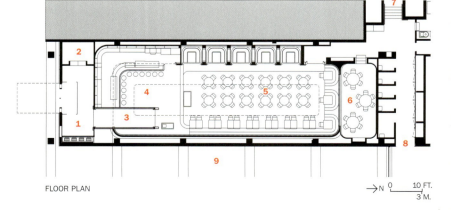

FLOOR PLAN

Wherever there's an architect, there's Architectural Record

哪里有建筑师，哪里就有《建筑实录》

世界精彩创新建筑＋
中国建筑特色及新闻

《建筑实录》中文版
每年三卷：
四月、八月、十二月

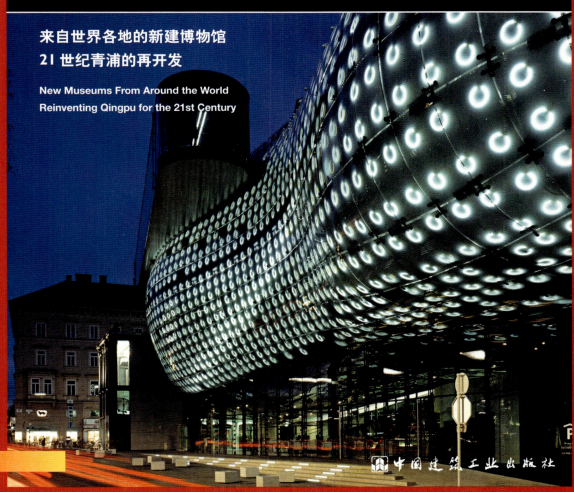

出版发行：中国建筑工业出版社
国内订价：29元

订购方式：
- 到全国各地中国建筑工业出版社销售网点购买，书店名称、地址请查询 www.china-abp.com.cn 销售商园地
- 网上书店：www.china-building.com.cn
- 邮购：电话 010-51986777 转 19、20、21、22，传真转 18

如果您对销售还有其他问题和建议，
欢迎直接与出版社发行部联系：
- 电话（传真）：010-68325420
- 地址：北京百万庄中国建筑工业出版社发行部
- 邮编：100037

The McGraw·Hill Companies

旅馆介绍译文

W饭店，纽约
亚布·布什伯格
P22

从好莱坞星球餐厅（Planet Hollywood）到位于时代广场W饭店（W Hotel）的时髦区域有一段很长的距离。52层高的塔楼最初计划作为一家狂热、大众化的好莱坞星球主题餐厅的连锁店。然而，随着建造期间业主的改变，流行文化的场景在一个更优雅的领域前变得黯然失色。亚布-布什伯格（Yabu Pushelberg）室内设计公司对酒店内部重新装修后，W酒店宁静的风景几乎是漂浮在这个号称"世界十字路口"的喧闹街头。

摒弃了低俗的大众趣味，如今酒店的内部装潢迎合了更优雅精致的品位。如同上了漆的便当饭盒，天然的色彩和纹理营造出禅宗般的气质，趣味的盛宴在交织缠绕的空间中展开。这是一个汇集了各式情调、功能与材料的取样器——它将零散的片段整合为一个和谐的整体。

当好莱坞星球餐厅在1999年提出破产保护的申请后，喜达屋饭店及度假村集团（Starwood Hotels & Resorts Worldwide）介入进来，力图将建筑改造为旗下W品牌的旗舰店。W品牌是在美国国内拥有16家连锁的都市酒店，专门面向钟情于设计的顾客。喜达屋抛开了以前的装修说明书，委托来自加拿大多伦多的建筑师乔治·亚布（George Yabu）和格伦·布什伯格（Glenn Pushelberg）工作组重新设计公共空间和客房。正如布什伯格所说的，给人"一片感官超负荷的剧场区中的绿洲"。

设计者必须应对建筑所面临的种种限制。四四方方的公共区域铺设着紧密的地板，叠摞在下面的楼层上，几乎没有连接和过渡的空间。在街道标高上，有独立的入口分别通向大厅、餐厅和一家地下的夜总会，进一步打断了空间的连续性。餐厅和夜总会已经租赁给外来的经营者，这更增加了问题的复杂性，因为这就必须给更多的业主阅方案。

"饭店的公共空间像谜题的片断一般组合在一起，"布什伯格说。顾客也许只逗留在某一个区域，餐厅或者四个酒吧中的一个，然而一个登记了的旅客就可能需要一整套的体验来充实呆在这儿的整个星期。在设计调整基础设施的时候，布什伯格解释道："我们将探索空间划分的新概念，即引导客人穿过体验一系列各式各样的、伸展演变的空间。"

尽管这个项目中代表民粹主义的好莱坞肖像已经被抛弃，"但我们明白在'时代广场'这个城市最繁忙的旅游区，'大众'才是它的核心。"布什伯格说。"而对备用品、线型装饰和围合空间不玩弄花哨，平实运用，则意味着用平易近人而非对抗的现代气质来吸引每一个人。"

正如亚布·布什伯格最近在纽约对蒂法尼（Tiffany）和伯格多夫·古德曼（Bergdorf Goodman）精品店的重新装修一样，"W"和它们共同展现出一连串惊人的视觉瞬间。进入位于第47大街的电梯厅，客人们便跨入一个戏剧性的门槛。流水在玻璃板后面顺着顶棚与侧墙缓缓流淌，映衬出休息厅的背景，紧凑的空间沉浸在波光闪动与水声潺潺之中。黑色的水磨石地面和不锈钢电梯门切削出冷峻的棱角。

到达第七层的接待大厅，电梯门正对着一个被称作起居室的休息厅。特大号的灯光盒子在白色皮软垫的长椅上方盘旋而上，穿入阁楼式的空间。白漆橡木地板、水磨石平台、丙烯酸立方体以及石膏和树脂的墙体——这是单色的调色板，偏偏有彩虹般缤纷的光线自丙烯酸屏罩后的巨大壁龛中投射下来，萦绕着敞阔的厅堂，将那单色衬得分外抢眼。由于现有的窗户可以俯瞰邻座的办公楼，设计者有意模糊了窗口来追求一种好似蚕茧内的宁静气氛。从外围粗犷的木质椅子到内部吧台近旁的长绒豪华沙发和座椅的选择引发了不同情调的交织。

起居室的西边，登记处设在一个昏暗的直线形"容器"内。黑色水磨石地面深色的面板和起肋镜面玻璃的插片把登记处衬托得棱廓分明。在划分出的柜台后，一块狂热涂绘的红色帆布使墙面有了喷漆的效果。深处的角落里，礼品店就如一个亮晃晃的白色立方体般闪现。

朝向百老汇大街，两层的"蓝鳍（Blue Fin）"餐厅容纳了两个鸡尾酒吧和一家寿司店，却感觉不出有400座的规模。上过黑色油漆的黑色松木板条被串在一起形成具有雕刻感的分割，变成了更有亲和力的图案的屏风。墙面色彩融合着琥珀、木炭与大红的调子，在清漆木桌上洒下柔和的光，沙里宁（Saarinen）椅子环绕在周围。沿着水磨石楼梯，混凝土墙面塑出波浪形的纹理，飘动着升至头顶上空的抽象鱼群。水的主题就这样被美妙地勾勒出来。

在地下室560m²的"威士忌"夜总会里，一连串的窗帘笼罩着坐席区和独立放映室，并配备了皮质的长沙发和凳子。一个老式的1970年代的舞池仿佛在向曾经设在这里的迪斯科舞厅致敬。充入了液体的嵌板在舞者经过时变幻出缤纷的色彩。

楼上共有509间客房，每一间都装修简约却不乏性感。家具椅在泛着烟熏炭色与烟草色的阴影里，靠着涂了暗蓝至沙黄渐变色阶的哑光墙面。一个连续的中性面——树脂与水磨石的混合——从入口的地板延伸至浴室，继而向上沿墙攀延至花洒。半透明的聚碳酸酯屏风充当了分隔者，切出浴室里的一个角落，模糊了公共与私密的空间界限。

就像亚布-布什伯格在W时代广场店里摆弄的空间一样，这些场景已经彻底改变了商业区连锁酒店的装饰标准，它把多种的惊喜都塞进了一个剥离分割后重新改装的容器之中。

本文原载于《建筑实录》，2002年9月
（段巍 译，蒋妙菲 校）

普赖斯大厦旅馆,俄克拉何马州
温迪·E·约瑟夫
P28

1956年2月9日,俄克拉何马州巴特尔斯维尔镇(Bartlesville)的居民成群结队地去参观弗兰克·L·赖特(Frank Lloyd Wright)为他们镇中心设计的惊人的建筑作品:飞扬的遮阳板、插入式的阳台、没有直角的轮廓。如何理解这座建筑?当地的居民把它当作一艘太空船,一个罩子似的装饰品,甚至是"普赖斯干的荒唐事",他们把矛头直指哈罗德·C·普赖斯(Harold C. Price),因为是他把项目委托给了赖特。

"人们认为这座建筑是疯狂的,因为它根本没有墙体",普赖斯的儿媳卡罗琳(Carolyn)回忆说,"它看起来就像是鱼的脊骨。但是经过这么多年,人们开始喜欢它了。你可能会说这座建筑已经从疯狂的怪诞发展成了一个精彩的奇迹。"

如今大楼已经连同一个由纽约建筑师温迪·E·约瑟夫(Wendy Evans Joseph, AIA)设计的拥有21间客房的精品旅馆(boutique hotel)一起被改造成了艺术中心。巴特尔斯维尔镇的居民希望这座昔日的奇迹魅力永存。自从康菲石油公司(Conoco/Phillips)总部搬走以后,这个镇就一直很混乱,它需要新的动力。普赖斯的大小旅馆应该成为提升这个城市活力的新动力。难道还有什么其他地方能让你既观摩赖特设计的摩天大楼,又在里面住上一晚吗?

因为哈罗·C·普赖斯在做管道生意时赚了大钱,所以他希望为家乡做点特别的贡献,也许是建个新的城市标志[见《建筑实录》,1956年2月,第153页]。因此1952年,赖特提出一个风车方案,一栋19层高的混凝土现浇大楼像树枝一样从中心悬挑出的楼板支撑着一个由办公室、公寓及商店组成的综合体。尽管有着不对称的平面以及更为复杂的功能,这个设计仍然是赖特1929年在Bowerie项目中未建成的圣·马克斯(St. Marks)方案的精练推祟。塔里埃森(Taliesin)的学员费伊·琼斯(Fay Jones)也参与了设计及制图工作。

"直根(Tap root)"的构思不断重复出现在赖特的作品中,包括这座建在偏远的俄克拉何马石油城中的功能异乎寻常繁杂的建筑。巴特尔斯维尔大厦是对赖特有关工作生活整合、技术解放力量以及天性和情感优先等观点的建筑性的总结。与其称之为建筑作品,倒不如说它是赖特的社会宣言。

哈罗德·C·普赖斯对赖特的构想起初是反对的,他告诉赖特,自己一直认为应该建的是一座两到三层的建筑。"那太浪费土地了。"建筑师回答。"但是对于这样一个小镇来说这简直就是个庞然大物。"普赖斯回应道。"一点也不,"赖特答道,"我只不过是把它竖了起来。"

于是项目如期进行,赖特最终实现了他所希望的每一件事,甚至在楼里设置了俄克拉何马州公共服务公司的分部。(艺术中心的主展厅就是过去的缴费窗口。)同样,他也给普赖斯造成了100万美元的预算短缺。

不过,曾经的沮丧终于变成了喜悦。普赖斯将大厦登在了公司杂志《Tie-In》的封面上,并且不错过任何一个展示其风采的机会。据说,员工们都喜欢这个建筑。"你不会被限制在一个小房间中,"公司管道涂层部的负责人比尔·克里(Bill Cree)回忆说,"你能够起床,散步并看到天空和绿地。有着这样的景观,你能够想出许多好点子。"但是其他人可不这么认为。占据每层楼四分之一的复式公寓后来发觉不仅又小又热,而且非常昂贵,简直就像曼哈顿的公寓。因此普莱斯将大部分这样的房间变成了办公室,但由于有太多固定的三角形空间,这些办公室几乎无法办公。

普赖斯公司一直拥有这栋大楼,直到1981年,他才将其卖给菲利普石油公司。菲利普石油公司在其中工作了约20年,之后该大厦成为普赖斯大厦艺术中心——一个研究当代艺术及建筑的非赢利性学术组织。该中心2001年建立,踌躇满志地安排了一系列展览和公众活动,并希望借助旅馆和餐厅的收入为其筹建一个基金。同时,艺术中心还在努力筹集1500万美元用于建造扎哈·哈迪德(Zaha Hadid)设计的博物馆[见《建筑实录》,2003年7月,第30页]。这是一座低层建筑,像飞镖一样将大楼围绕起来,同时伸展出去与赖特的女婿威康·W·彼得斯(William Wesley Peters)设计的表演艺术中心相连。没准儿哈迪德大胆的几何形体将引发另一轮公众争议,从生疏到友好的循环又将再度上演。

温迪·E·约瑟夫设计的旅馆承担着与赖特的建筑进行精神对话的使命,但她的设计没有陷入模仿或草率,她将之延续,时有对比,但绝不复制。"这并

旅馆介绍译文

不容易。"她解释说,因为这座建筑极其特别,它有着强烈的几何感却又很难找出基本的元素,以至于我很迷惑是否应该抛下赖特的限制另寻出路,然而我发现我不得不与他在每个关键点上继续抗争下去。"

她严格地视其为室内设计工作。对于大楼,她没做什么结构性的调整,空间也保持了赖特设计的样子,但是却赋予了其不同的功能。18了间办公室和原来的套公寓都变成了客房(每晚125~250美元),第四套则改成了Copper餐馆。门厅连同赖特钟爱的诗人沃尔特·惠特曼(Walt Whitman)的语录一起被忠实地保留下来。此外还有顶层哈罗德·C·普赖斯的办公室,看起来就像是将时间停滞在了普赖斯离开的那一天。

在更新材质和技术时,约瑟夫成功地保留了赖特的设计灵魂,例如,她在客房和餐馆中大量使用铜,不过是以轻薄的管状丝网来代替厚重的绿铜板。类似地,地毯及装潢织物也使人回想起赖特的主题,但它们并没有复制任何一个具体的图案。赖特对普赖斯大厦的比喻——"一棵逃离拥挤森林的大树",它成为有着当代风格,甚至带着些许日式风格的客房壁饰的出发点。

赖特设计的桌椅运用了纯粹的形式,犹如微型的雕塑,而约瑟夫则用细长的枫木条突出建构,引发再创造。赖特的设计就像是锚固在地上,而约瑟夫的设计则看上去仿佛是漂浮在空中。因为大楼电梯与电话亭的大小相仿,单独的部件不得不被一个个地挂起并在现场装配。同样的方法赖特也曾用过。

位于大楼十五和十六层的复式餐厅Copper是视线的最高点,它起初是公寓。吧台呈现出柔长和缓的曲线,侧壁层叠的枫木是对赖特漩涡形古根海姆美术馆(Guggenheim)的回应,顶部则以铜皮饰面。餐桌的桌面是夹着铜丝网的双层玻璃,同样的铜网也应用在窗帘上,与建筑外部厚重的铜质遮阳板形成流动的呼应。餐厅好似一座三维雕塑,任意两个表面或视角都不尽相同。

花费210万美元的翻新让普赖斯大厦回到了原点——从新奇的焦点到先锋派的象征,到公司的弃儿,直到社区的标志。它的原型圣·马克斯是单纯的公寓,它当前的化身是更加时髦和更加瞬变的住所。"这是一个独一无二的向赖特精神致敬的机遇",艺术中心主管理查德·汤森(Richard Townsend)说,"同样这也是一个创造社区焦点的机会。这就是哈罗德·普赖斯当年想要的节镇广场。"同时,在这个过程中,也再次引起了公众对美国这个最浪漫、最特殊和最具强烈个人主义特色的建筑作品的关注。

本文原载于《建筑实录》,2003年7月
(董晓霞 译,蒋妙菲 校)

"Q!"旅馆,柏林
Graft
P36

在任何一个有格调的欧洲首府寻找奢侈品店铺,如果立即想到"设计师旅馆(design hotel)",那么离目标就不会太远了。在柏林,从夏奈尔(Chanel)、路易威登(Louis Vuitton)、卡地亚(Cartier)专店一路过来的拐角处,一座新开张的入口别致却没有招牌的旅店正印证了这一规则。不过对于那些熟门熟路的人,没有招牌的"Q!"并不难找。那种在这座城市里随处可见的窗户嵌在寻常的灰色墙面内,首先显示出"Q!"与其在这个紧邻库尔菲尔斯滕大道(Kufürstendamm,)

的租金昂贵的地段上的邻居们截然不同的姿态。从街上看去,半透明的白色窗帘掩蔽着旅馆,但是只需跨过玻璃前门,来宾就进入一片红色表皮包裹着的世界,看上去更像是在加利福尼亚州而不是新柏林。

从门厅进入休息室和餐厅,曲线型的红色漆布饰面从地板一直延伸到墙面。沙发和镶壁家具同样打上了建筑师的商标——Graft。这家年轻的位于洛杉矶柏林两地的公司拥有20名雇员,它的合伙人沃尔夫冈·普茨(Wolfgang Putz)、托马斯·威廉姆特(Thomas Willemeit)、拉尔斯·克吕克贝格(Lars Krückeberg)把得到这个项目更多地归功于"好莱坞姻缘",而非"德国出身"。旅馆的经营者沃尔夫冈·卢克(Wolfgang Loock)是在2002年看到一篇关于他们为演员布拉德·皮特(Brad Pitt)设计的好莱坞山工作室的文章后给他们打的电话。

到Graft接手这家旅馆(该公司的第一家旅馆)的室内设计时,尽管建筑还没有动工,但开发商早已选定了另一位建筑师设计这座7层的建筑。虽然Graft与那位建筑师对于这个项目的设计几乎没什么同感,但仍然得设法与之合作。在提议了一个不同的立面(未能实现)后,合作伙伴们承认这是个棘手的任务——既要在紧张的预算(大概100万欧元用于室内装修)下实施这个挑战性的项目,又要应付一个起初并不大认同他们理念的客户。除此以外,入口近旁现存的一部电梯和楼梯以及区区370m²的占地面积也是个不小的限制。

在整个3000m²的建筑中,Graft成功地塑造了一种风格统一的美,这种手法不可否认出自建筑师尼尔·德纳利(Neil Denari)的折叠平面的影响。尼尔·德纳利恰好是在普茨和克吕克贝格从南加利福尼亚州建筑学院(Southern California Institute of Architecture,简称SCI-Arc)毕业时走马上任学院院长的。在"Q!",连绵不断的流线型表面不仅包裹了首层的公共区域,而且还延伸至客房室内,色调也从火热的色系转变成清冷的色调。白色与橡木地板的烟灰色形成冷静的对比,让人从热情的喧嚣回复到淡定的宁静。在这儿,墙体与顶棚、台面融为一体。头顶曲线形的顶棚影影绰绰地印着克里斯蒂安·托马斯(Christian Thomas)的女人肖像摄影作品,目的是给这些地方一种普茨称之为"犹在茧中"的感觉。

精心设计的房间要追求包裹的平滑度,因而放弃了许多诸如门把手这样有着特定日常功能指向的构件。橱柜或者电灯开关亦非显而易见。如普茨所说,"我们希望来宾花一些时间自我定位。"虽然在新的设计师旅馆中,似乎只有考究的装潢材料才是合乎时宜的,但Graft在相当低的预算内精心设计,以精湛的工艺传达高品质的感觉,巧妙地迎合了这种审美时尚。比如在浴室中看起来像石板的材料其实是黑色的陶土,如此等等,不一而足。

朦胧的的顶棚照片可能会使人回想起让·努维尔(Jean Nouvel)在"瑞士卢塞恩市(Lucerne)设计师旅馆"[见《建筑实录》2001年5月,第238页]中对于电影图像与建筑更有说服力的整合,但Graft的建筑师声称他们是在其他地方找到的灵感。他们将自己的作品比作电影片断。然而普茨认为,柏林大部分的建筑更多地倾向于静态的图像。与布拉德·皮特一起工作的经验促使Graft用这些叙述性的语汇来思考建筑。无论在酒吧还是在那些浴缸与床并在一起的客房,建筑师像电影导演一样构思着这些空间,想像着客人们在室内外活动的种种场景。

"Q!"试图将菲利普·斯塔克(Philippe Starck)在伦敦圣马丁乡村旅馆(Saint Martins Lane hotel,简称SML)中的风格及设计思想带到柏林[见《建筑实录》2000年1月,第90页],或者不花一分钱带来伊恩·施拉格(Ian Schrager)的特征。虽然Graft的合伙人声称他们从未看过SML的室内,但看上去似乎不仅德纳利,连斯塔克和努韦尔的影响也已经某种程度地融入了"Q!"。旅馆的设计以及服务都给人一种独特的déjà-vu(法语,实际上从未经验过的事情却仿佛在某时某地经验过似的,一种"似曾相识"的感觉)印象,但或许Graft的设计中那种使人耳目一新的加利福尼亚州特色是它的开放与乐观,超越了合伙人称之为的"典型德国式怀疑论"。在如此紧凑的空间和有限的方式下,Graft的建筑师们没有否认可能性而拒绝这个项目,他们将当代设计中的平滑曲线融入到一个灰色的棱角生硬的"柏林盒子"里,如此而已,并没有使用什么其他的小伎俩。

本文原载于《建筑实录》,2004年9月
(蔡瑜 译)

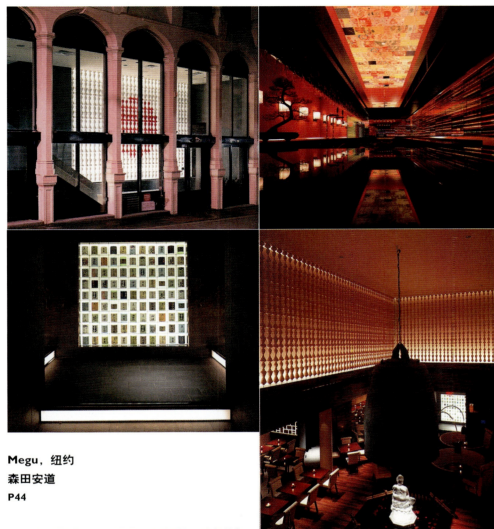

Megu,纽约
森田安道
P44

2005年3月,Megu在翠贝卡区的开业在纽约餐饮业引起了一时轰动,该餐馆无论是食物、服务、设计或是价格都有着传奇色彩,仿佛为银幕而生。倘若第22街上罗科·迪斯皮利托(Rocco DiSpirito)开业1年的餐馆有着电视般写实的话,Megu则展现出电影式的彩色幻想。

如果日式餐馆在你的印象中就是亚麻色的木器和优雅的小型寿司吧,那么Megu带来的将是一种震撼。在这家两层的餐馆里恐怕难得找到一方静土,从一进门服务生高喊的"欢迎光临",到大胆的用色和对材料不拘一格的混用,所有这一切与传统日式餐馆的宁静大相径庭。也许可以称之为现代日式巴洛克(Modern Japanese Baroque)。餐馆的设计风格与那些奢华的菜肴倒是相得益彰——桌上哧哧作响的滚烫岩石上烧着神户牛排,手持着炽热铁棍的侍者让浸着托洛红酒的鲑鱼拌着芥茉酱慕丝在眼前慢慢酥软。

35岁的企业家金井岛(Koji Imai)是Megu的创始人,他在日本拥有30家餐馆。金井以Megu首次冲击美国餐饮业市场,他也希望这是他在纽约,乃至美国其他城市经营餐饮连锁店的起点。金井聘请了来自大阪的年轻设计师森田安道(Yasumichi Morita)作为他美国餐饮旗舰店设计队的领军人物。早在2002年,他们就在东京神水户(Tokyo's Shinjuku)开业的麦孟(Maimon)餐馆项目中一起合作过。

项目计划 作为一家开在曼哈顿的新生代大型餐馆,Megu占地1300m²。除了200座的主餐厅以外,它还包括一间朱红色调的"和服吧(Kimono Bar)",一间能俯瞰餐厅的"皇家休闲室(Imperial Lounge)",一间原先设计为吸烟室的小型VIP室,一间寿司吧和一间靠近厨房的私人用餐室。餐馆设在一幢坚固的19世纪建筑的底层及地下层。

解决方案 "Megu实在太大,因此我们把它设计成一系列不同的场景,"森田解释说。场景始于人行道,客人们可以看见休息厅那面背光照明的嵌花式外墙,墙面正中饰有一个日本国旗意象的红色太阳图

旅馆介绍译文

案,走近一看才发现这面墙其实是由瓷质清酒瓶和饭碗垒成的圆柱体构成。就像精心制作的电影片头,入口的墙面为后面的场景埋下了伏笔。对日本文化象征的重新诠释以及对于古老材料醒目的新式运用,产生了Megu的主题,展示了Megu的非凡想像力。

走过瓷瓶、瓷碗组成的墙,第一个充满戏剧性的场景发生在酒吧间。酒吧间的两面墙上排满了一卷一卷的各色和服面料,头顶被照得通明的长方型吊顶内铺满一方一方的和服面料,看上去颇有几分类似传统的被褥。森田利用镜子和屋内鲜亮的中国红来增强和服面料的奢华效果,在顾客点酒前已渲染出一种流光溢彩、目眩神迷的氛围。

设计师巧妙地将安静和喧嚣的场景轮流交替布置,例如酒吧后的休息室,室内米色的梅塞德斯(mercedes)皮革和弯曲的长沙发营造出慵懒闲适的情调。不仅如此,他还精心安排了客人们穿梭于整个餐馆时的体验,比如,引导客人们走下一段双向的狭窄石阶,当顾客到达他们的餐桌时,会觉得2层高的餐厅看起来比实际的更大。

不论在何处,森田都会用他独创的方式处理司空见惯的材料。在通往休息室的路上,顾客会经过一面用贴在玻璃上的日式火柴盒贴装饰的墙。在楼梯平台上,他们会惊叹于那些贴在曲面塑胶底座上的清酒标签构成的格栅在背后灯光的映照下居然宛如一幅精美的艺术画。在餐厅东侧墙面,设计师将竹席拼成棋盘格图案;在西侧墙面,则将薄薄的矩形条石粘在玻璃上,形成仿佛旧式砖墙的纹理。

餐厅中央悬挂着一个重达320kg的大钟,它仿造日本奈良(Nara)某寺庙中一个更为巨大的铜钟铸造而成。钟下一尊冰雕佛像在渐渐消融,没入莲花座下漂浮着芙蓉花瓣的水池。在这个宽敞的餐厅内,最吸人眼球的差不多就是这些几乎称得上是艺术品的大钟和佛像了。

评论 森田和他的设计组超越了那种以日本传统的艺妓与武士图像来表达日本文化的陈词滥调,将日本文化转译为纽约人能够理解的建筑语言。在这个设计大胆、新奇,富有触感的餐厅里,前来用餐的人们如同置身于电影中,体验着场景随着空间变换的奇幻旅程。"来到Megu不仅是为了满足食欲,"森田说,"更是为了享受。"事实上也是如此。

每当人们乘飞机环游世界时,一个大洲接着一个大洲的画面从眼底跳过,Megu则通过一位日本艺术家向美国观众提供了一个对于现代日本高效而充满活力的诠释。这个诠释可信吗?它会带来什么不一样吗?这就是娱乐业。

本文原载于《建筑实录》,2004年7月
(缪诗文 译,蔡瑜 校)

帕提娜餐厅,洛杉矶
哈吉·贝尔茨伯格
P50

在迪斯尼音乐厅(Walt Disney Concert Hall)内设计餐厅和咖啡厅绝对是一项充满潜在风险的项目。新的设计既不能在弗兰克·盖里(Frank Gehry)的大作前抢尽风头,又不能甘拜下风。其中可能出现的最危险的失误就是将迪斯尼音乐厅这个洛杉矶星光大道(Los Angeles's Grand Avenue)的角落变成"伪盖里地带"(ersatz Gehryland)。这个极富魅力与挑战的项目引得圣莫尼卡(Santa Monica)当地建筑师们争相请缨,哈吉·贝尔茨伯格(Hagy Belzberg)最后被选中担此重任。"我们应该尊重盖里的设计,但我们必须有自己的特点",贝尔茨伯格说。

项目计划 贝尔茨伯格之前设计的餐馆和住宅以材料丰富、空间活跃见长,而他现在必须在迪斯尼音乐厅既定的外壳内展开设计。在这里,他接手的是平板玻璃立面内一条凹进的、几乎就是从属于盖里的流线形体的带形空间。正立面现有的主入口成为贝尔茨伯格设计这个460m²的底层餐厅和370m²的咖啡厅的起点。

解决方案 竣工后,餐厅的主入口开在吧台前方的等待区。红酒瓶整齐地排列在半透明的玻璃墙后,瓶口的软木塞齐刷刷地朝向前方,在背光照明中形成一种别致的图案。入口的一侧是一间小型私人派对房。半透明的窗帘遮住沿街的视线,而透过落地窗可以瞥见厨房;入口的另一侧是主餐厅,室内呈现出深沉、暗哑而含混的色调。四方的玻璃推拉墙上反映着街对面洛杉矶当代艺术博物馆(Los Angeles Museum of Contemporary Art)投射来的艺术气息。

迪斯尼音乐厅早在2002年9月竣工揭幕之前就成为一幅历史图像。在贝尔茨伯格看来,用电脑技术加强的隐喻被证明是对抗盖里这幢神话般、有着潜在压倒性的建筑的最好方式。对于贝尔茨伯格的这个高档餐厅(甚至在整个工程建设期间,其租赁期依然尚未决定)来说,剧场的帷幕是其设计的灵感源泉。毕竟,这里是开演之前和表演结束后人们最活跃的活动地点。表演大厅内帷幕的缺失——它用伸出式的舞台取代了传统的舞台——给这一隐喻以令人兴奋的切入点。

无论多么柔软的幕帘,都是在它们被拉紧并悬垂下来以后才显得生动活跃。这个动态凝固的概念激发了贝尔茨伯格通过设计这样一种效果来衬托迪斯尼音乐厅立面含蓄的动态或者航行般的形态。

贝尔茨伯格同样在设计过程中充分运用电脑来回应盖里非凡的建筑。盖里很好地运用了航空学程序CATIA将迪斯尼音乐厅起初用卡纸手工制作的模型变成三维实体模型。贝尔茨伯格选择用Z形(form-Z)软件来支持电脑生成优于固定形式的形态。

贝尔茨伯格在屏幕上设计出波纹并使之定形,再将指令传送到铣床上,从坚硬的胡桃木夹板中雕刻出"帷幕",从而引起伸展的视觉效果。他将这些"窗帘"挂在餐馆的大部分室内,通过滑动的不对称的结构界定出开放空间。波纹与胡桃木本身的纹理和规格材的模数格格不入。巨浪中翻滚的"帆"转化成微波起伏的"布",这些板材将盖里的大手笔进行了迷你型的转述。理论上,贝尔茨伯格则迫使汹涌的大尺度形式转变成装饰尺度,在盖里的建筑内不安分地搅动着。

顶棚上的波纹是最深的,在那里,胶合板条弯曲起伏的表面被嵌在凹槽缝隙里的冷阴级灯照得通明。顶棚内条纹灯光的排布形成大波浪,成为波纹面板与盖里的宏伟曲线之间的过渡尺度。

评论 贝尔茨伯格设计了得体的室内,在形式、空间和技术上都与迪斯尼音乐厅协调一致,但绝不是顺从。由电脑创造出的曲线与直角的混合形将建筑体宏大的气势转化到小尺度。有别于盖里的隐喻,贝尔茨伯格建立了他自己的隐喻。而两种途径的一致性提供了一种天衣无缝的渐变。帕提娜餐厅的设计要的是恭敬地区别于迪斯尼音乐厅,而不是彼此显而易见的对抗。

本文原载于《建筑实录》,2003年12月
(缪诗文 译,蔡瑜 校)

66号餐厅,纽约
理查德·迈耶
P56

自66号餐厅于2002年春开业以来,让-乔治·瓦格里奇(Jean-Georges Vongerichten,国际著名烹饪大师)的海派烹饪风格和理查德·迈耶(Richard Meier,FAIA)的极简主义风格室内设计,使它获得了餐厅主人渴求的热切关注。两位业主瓦格里奇和菲儿·苏亚雷斯(Phil Suarez)都是厨师,也是迈耶在西村(West Village)设计的佩里街公寓(Perry Street Apartment)的投资人之一。因此,尽管迈耶此前只设计过一家餐馆——位于洛杉矶的格蒂中心(Getty Center),来自其他餐厅的竞争在这个小山顶显得微不足道——苏亚雷斯也没有犹豫。他说,"我们相信迈耶一定能为瓦格里奇的烹调风格提供令人兴奋的恰如其分的环境。"[尽管如此,瓦格里奇的餐馆顾问马克·施特希-诺瓦克(Mark Stech-Novak)还是在现场布置厨房以及提出诸如此类的建议;同时,照明顾问L'Observatoire International公司确保了室内照明,使迈耶著名的"白色派"格调仍能给人以温暖而亲切的感觉。]

项目计划 瓦格里奇与苏亚雷斯此前已经营了许多住宅区内的餐厅(如Jean Georges、Vong、JoJo),因此他们二人决定在纽约翠贝卡区(Tribeca,纽约艺术家社区的中心区,著名富人区之一)的纺织品大厦(Textile Building)底层开设一个柔和的古典风格餐厅。有着百年历史的纺织品大厦由亨利·哈登伯格(Henry Hardenbergh)于1901年设计。哈登伯格也是纽约著名的广场酒店(Plaza Hotel)与达科他公寓(Dakota apartments)的设计者。这座地标建筑的不远处就是奥迪恩(Odeon)。它是位于市中心区最早的带有艺术家风格的餐厅之一,于25年前开始营业。在此期间,翠贝卡区成为工作室相对集中的艺术家社区,有统一的公寓式管理服务,迎合了那些喜欢轻松生活方式的艺术家们。

因此,迈耶认为餐厅应该是开敞和明亮的。他说,"我希望人们能产生如同身处更大空间的舒适感。""而且",迈耶补充说,"我认为进餐空间不应该有层次之分。无论坐在何处,你都能感觉到这是最尊贵的位置。"

解决方案 迈耶将矩形空间分割成三个主要部分,围绕着一个用3.6m高的曲面毛玻璃墙限定出的中央入口门厅。通高的毛玻璃面板分隔出不同的区域,固定的不锈钢网再将空间细分为隔间,内设木底皮面的长沙发。

入口门厅的正后方,一个13m长、40座的公共餐桌成为餐厅的定位点。一排红色的丝质幌子从吸声板吊顶的槽内悬挂下来,渲染出一种戏剧性的氛围。公共餐桌后面的吧台隐蔽在一片毛玻璃墙后,酒保活动的身影与酒瓶排列的轮廓在毛玻璃墙上若隐若现。

在进餐区,透过四个养着彩色鱼儿的玻璃水族箱可以瞥见厨房。这间布置得干净整洁、井井有条的厨房主要用于最后阶段的烹饪,卤素灯用来防止刺目的眩光进入餐厅。(另一个用于初加工的厨房设置在地下室。)

评论 阁楼翻新(改用喷漆的锚固钢柱支撑)与精心设计的小餐厅的结合,显示了迈耶出名的对细节与工艺的注重。对半透明的玻璃隔断与闪烁的不锈钢网眼隔栅两种材质的精巧运用,实现了渴望达到的开敞性与私密性的结合。同时,这一切也使餐厅更加精致典雅。在柔和的环境照明(包括夜晚的烛光)和点缀其间的些许色彩(如红色的幌子、彩色的游鱼)的配合下,白色的主色调非常出彩。

当初设计长长的公共餐桌似乎只是一时的时尚需求,主要是针对手机文化——促进了互不相识的人们接听私人电话,到如今看来效果还不错。昂贵的价格以及由此支持的高水准的烹调质量,使66号就像一块磁石,吸引着慕名前来、甚至临时寻找用餐地点的人们。66号能够与价位稍低一些的奥迪恩餐厅的长盛不衰及其酒吧的开放活跃场面相匹敌吗?我们拭目以待。

本文原载于《建筑实录》,2003年12月
(蔡瑜 译)

科略斯特里餐馆,瑞士格施塔德
帕特里克·茹安
P60

在通往瑞士山区旅游胜地格施塔德(Gstaad)的主干道旁,一座有着300年历史的瑞士传统小木屋如今成为科略斯特里(Chlösterli)餐馆。餐馆不仅融合了传统与现代风格,还带着些许幽默。木屋最初由修道士鲁热蒙·阿比(Rougemont Abbey)主持建造,曾被用作餐馆和比萨店,后来被摩纳哥开发商米歇尔·帕斯托(Michel Pasto)买下。帕斯托和主厨阿兰·杜卡斯(Alain Ducasse)邀请巴黎设计师帕特里克·茹安(Patrick Jouin)给这幢深色的木构建筑带来新生。茹安现年37岁,曾与杜卡斯在巴黎的Plaza Athenée餐馆和纽约的Mix有过合作。在1998年创立自己的公司之前,他还为菲利普·斯塔克(Philippe Starck)主管家具及产品设计。

面对这幢村庄里现存的最古老的木构建筑,茹安恪守保护原则,清理并修复了小木屋的立面。对外观最显著的改动是新增了一个150m² 的夏日露餐平台。露台由绿柄桑木和混凝土搭建而成,四周木栏环绕,板条布局恰如其分,使山林绿野、田园美景尽收眼底。

项目计划 杜卡斯要求设置两个餐厅 一个位于底层,是具有瑞士传统风格的用餐场所;另一个是在楼上的Spoon des Neiges——全球七家Spoon

旅馆介绍译文

餐厅之一〔茹安在圣特罗佩（Saint Tropez）设计了Spoon Byblos，于2002年开业〕。杜卡斯在巴黎、摩纳哥、纽约以及法国的城堡酒店都经营了一系列备受称赞的餐馆，是位标准的大忙人。

科略斯特里每个餐厅都独立拥有一间200m²的厨房，用于供应不足100m²的就餐区。由于客户定位是有钱人，一层餐厅还设置了一个80m²的迪斯科舞厅。

解决方案 茹安以木屋室内的深色木材为立足点，将现代性及对瑞士传统半戏谑的尊重用一种意想不到的方法融合在一起应用到设计中。就餐者从低调的临街入口进入，石板地面与橡木墙板奠定了底层餐厅的基调，在这里就餐几乎感受不到什么现代特征。茹安只在其中稍稍改变了一些传统做法，为橡木椅加上马鞍式的皮面（茹安设计了该项目中所有的家具及灯饰。）。

根据普通阿尔卑斯山居民的感受，两层高的迪斯科舞厅是最震撼的。地面用透光树脂板取代苏格兰石板，通过树脂板下面的电子显示系统产生地面振动，并打光改变空间颜色。5m高的玻璃隔墙将迪斯科舞厅与厨房隔开，同时也是一个庞大的透明葡萄酒架。桌子的设计灵感显然来自古老的酒桶，而木椅则是对其本土原型戏谑的变形。就这样，茹安在一个传统的农场里进行了国际式的混合，不经意地创造了一种不调和的共存。

两段狭窄的楼梯唤回人们对木屋乡村本色的记忆。拾级而上，就餐者进入二楼的Spoon，这里一反楼下的传统乡土气息，时髦的现代美学大行其道。在酒吧，因为当地法规禁止明火，一个等离子屏幕仿造"壁炉"的样子，闪烁着跳跃的"火焰"。钢骨皮面的座椅给这层楼带来更雅致的氛围。在一个绰号"水箱"的私人就餐区，就餐者可以透过一面通高的玻璃墙俯视迪斯科舞池。茹安在二层用完全现代的设计语汇，狡黠地完成了从瑞士往昔的传统文化到格施塔德现今的旅游文化的过渡。

评论 在一幢18世纪修道士建造的房子里设计时髦的餐厅，似乎多少有点儿讽刺意味。但茹安并不否认或者掩盖这个事实，而是借此作为设计的手段。他没有抹杀过去，而是利用过去，由此创造了一个优美而诙谐的环境，让就餐者向着一个日益现代的场景展开空间之旅。考虑到一些极端的情况，这种过渡没有招致因审美差异而引起的非议也确非易事。帕特里克·茹安以冷静的炫耀树立起独特的风格，并在这一过程中架设了一座从陈旧木构到迪斯科舞厅的跨越300年时间长河的桥梁。

本文原载于《建筑实录》，2004年7月

（董晓霞 译，蒋妙菲 校）

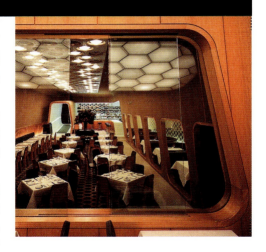

利华大厦餐馆，纽约
马克·纽森
P64

作为纽约公园大道上的现代图像（Modern icon），利华大厦（Lever House）的价值早在1983年被城市标志性建筑保护委员会（Landmarks Preservation Commission）指定为地标建筑时就已经确认，尽管当时这个SOM（Skidmore, Owings & Merrill）的作品仅仅拥有31年的历史（绝对算不上是一个古董）。截止到2002年它的50周年纪念时，对这座建筑的全面整修已接近尾声〔见《建筑实录》，2003年3月，第122页〕。但是赋予它新生所必需的绝不只是崭新的大堂家具和幕墙，对业主佛雷德里克·罗朗工作室（RFR）来说，关键的问题是如何激发底层空间的新活力，这里以前仅是作为会议室和利华兄弟（Lever Brothers）公司的仓库来使用的。

项目计划 利华大厦餐馆于8月份开张，餐馆主人约翰·麦克唐纳（John McDonald）和乔希·皮卡德（Josh Pickard）以此进驻纽约餐饮界。600m²的可用空间实际上是在地下，而且没有自然采光，仅在建筑南侧的第53街上设一个直接入口。因为地标法不允许在建筑上出现过分夸张的标志，所以餐馆的临街门面不能太招摇。马克·纽森（Marc Newson）承接了该项目，这位定居巴黎的澳大利亚设计师因其曲线型的设计而出名，他的设计涉及自行车、椅子、机舱内部和"你周围的一切物品"。这种曲线型设计实际上是由《Wallpaper》杂志流行起来的怀旧的现代派风潮。

解决方案 在将近3年的时间里，纽森设计出一连串的六边形和曲面，这些都是既复古（呼应了1950年代的母题）又流行的元素。纽森与他的顾问建筑师塞巴斯蒂安·西格斯（Sébastien Segers）一道，在无窗的空间中创造出窗户。他在房间的一侧砌了一面内墙，挖出巨大的曲线型洞口，看起来酷似旅客车厢上的窗。用餐者穿过洞口坐在弯曲的长沙发上，可以回望地坪略低15cm的大厅内的人群，但是感觉上会显得更远些。在远端的另一面墙上开有一个几乎与主餐厅同样宽的窗口，界定出22座的私人餐厅。窗上装有可滑动的透明玻璃，关闭时达到隔声的效果，但不阻隔视线，当然也没有什么私密性，因此这儿的用餐者就像是一直处在舞台上。这是个基于宴会剧场概念的手法。

纽森在酒吧和餐厅里用了再普通不过的材料：白色的可丽耐（Corian）、粗糙的灰黄色石膏、浅色橡木、时髦的镜面玻璃。与第53街垂直的纯黑色主入口突显了材料的轻盈与纽森赋予室内的曲线。衣帽间也在入口处，其后一条全黑的走廊通向全黑的洗手间（包括设备及所有一切都是黑色的）。视觉的暂时黯淡掩饰了对空间的密集使用带来的弊端，但更重要的是，黑暗强化了通向明亮的中央空间的感受。六边形的蜂巢图案无所不在——脚下的地毯、头上的格子顶棚、私人餐厅的照明，甚至是吧台的背景，都着实增加了几何情趣的视觉效果。

纽森继续为用餐者下一步的行为做好准备：一条白色可丽耐的曲面隧道灯光璀璨，从门厅倾斜延伸至地下主餐厅，就像运送宇航员一样运送用餐者进入另一种空间尺度。在这里，用餐者可以享用主厨丹·西尔弗曼（Dan Silverman）呈上的佳肴，几乎意识不到是在地下室。

评论 利华大厦餐馆于2003年8月开业。9月里一个星期二的傍晚，餐厅里人声鼎沸，据说没有预定的客人需要为一个座位等上足足两个小时，这真是一周中最漫长的夜晚了。蜂拥而至的人们是为了追求美食还是情调？

龙虾天麸罗（日本菜肴）和烤野生鲑鱼的确味道鲜美，但也并不比一个街区之外的由菲利普·约翰逊（Philip Johnson）设计的久负盛名的四季餐厅（Four Seasons Restaurant）里的菜肴更令人难忘。那个巨头工作餐会的据点按理应该是利华大厦的榜样，但是两者的装饰格调截然不同。与约翰逊的设计流露出的持久的优雅相比，纽森的空间更多流露出的是时髦。在细微精致上的缺失似乎注定了其流行的时间是不会长久的。

本文原载于《建筑实录》，2003年12月

（段巍 译，蒋妙菲 校）